THE SCRAMJET ENGINE

PROCESSES AND CHARACTERISTICS

The renewed interest in high-speed propulsion has led to increased activity in the development of the supersonic combustion ramjet engine for hypersonic flight applications. In this flight regime, the scramjet engine's specific thrust exceeds that of other propulsion systems. This book, written by a leading researcher, describes the processes and characteristics of the scramjet engine in a unified manner, reviewing both theoretical and experimental research. The focus is on the phenomena that dictate the thermoaerodynamic processes encountered in the scramjet engine, including component analyses and flow-path considerations; fundamental theoretical topics related to internal flow with chemical reactions and nonequilibrium effects, high-temperature gas dynamics, and hypersonic effects are included. Cycle and component analyses are further described, followed by flow-path examination. Finally, the book reviews the current experimental and theoretical capabilities and describes ground-testing facilities and computational fluid dynamic facilities developed to date for the study of time-accurate, high-temperature aerodynamics.

After completing his Ph.D. at the University of Virginia in 1991, Corin Segal took a teaching position at the University of Florida in the Mechanical and Aerospace Department, where he now leads research in the Combustion and Propulsion Laboratory. Prior to his graduate studies, Dr. Segal spent more than 11 years in the aerospace industry as a senior aerodynamicist and project manager and as a leader of the technical bureau. His current research at the University of Florida covers a range of topics, including mixing and combustion in high-speed flows, supercritical mixing, high-pressure combustion, and cavitation. Results of his group's research have appeared in national and international publications. Dr. Segal is an associate editor of the *AIAA Journal of Propulsion and Power* and an associate Fellow of AIAA.

Cambridge Aerospace Series

Editors

Wei Shyy and Michael J. Rycroft

The Scramjet Engine

PROCESSES AND CHARACTERISTICS

Corin Segal

University of Florida

CAMBRIDGE
UNIVERSITY PRESS

CAMBRIDGE UNIVERSITY PRESS
Cambridge, New York, Melbourne, Madrid, Cape Town,
Singapore, São Paulo, Delhi, Tokyo, Mexico City

Cambridge University Press
32 Avenue of the Americas, New York, NY 10013-2473, USA

www.cambridge.org
Information on this title: www.cambridge.org/9781107402522

First published 2009
First paperback edition 2011

A catalog record for this publication is available from the British Library

Library of Congress Cataloging in Publication data

Segal, Corin, 1952–
The scramjet engine : processes and characteristics / Corin Segal.
  p.   cm. – (Cambridge aerospace series)
Includes bibliographical references and index.
ISBN 978-0-521-83815-3 hardback
1. Airplanes – Scramjet engines.   I. Title.   II. Series.
TL709.3.S37S44   2009
629.134′353 – dc22   2008053025

ISBN 978-0-521-83815-3 Hardback
ISBN 978-1-107-40252-2 Paperback

Contents

Preface

There is, justifiably, a great interest currently in the study and development of scramjet engines for hypersonic flight applications. A major impetus is the potential to reduce space accessibility costs by use of vehicles that use air-breathing propulsion from takeoff to the edges of the atmosphere; defense applications for hypersonic flight within the atmosphere are also of considerable interest. In the hypersonic flight regime, commonly considered to begin when velocities exceed Mach 6, the scramjet engine's specific thrust surpasses that of any other propulsion system. Subsonic combustion, which technologically is easier to manage with the current knowledge, would be associated, in the hypersonic regime, with high stagnation temperatures that would lead to unacceptable dissociation levels, and hence an inability to materialize the energy rise expected through chemical reactions. Additional thermal and structural considerations preclude the use of other air-breathing propulsion systems at these flight velocities.

In the late 1950s, when scramjet research began, the development of this type of engine proceeded with varying degrees of intensity, as the national interests of the times drove investment levels. The past decade has seen increased enthusiasm in all sectors because of the expansion of government-funded scramjet research and numerous national and international collaborations, and development has been buoyed by significant scientific and technological progress. Major activities at the national level and international collaborations exist in Europe, including Russia, and in Japan, Australia, and the United States. This increased activity produced a great deal of knowledge; yet, much of the information accumulated over the years through various programs lies in the classified or proprietary category, and it is subject to limited availability. Several printed compilations have brought together major aspects of the international scientific and technical developments in this field. In most cases these volumes consist of contributed articles that summarize research and program results, including component technologies, national programs, and theories resulting from these efforts. The last major effort in this direction was the volume edited by E. T. Curran and S. N. B. Murthy in 2000 that completed a set of three volumes, started in 1990, dedicated to the subject. As the editors pointed out, only the last volume in the series was dedicated specifically to scramjet propulsion; it included contributions from researchers worldwide with updated reviews of national programs and their results.

This book is intended to offer the reader an introduction to the study of scramjet propulsion, including careful definitions of terms and a unified description of the processes and characteristics of the scramjet engine. This book reviews the major knowledge base that has been accumulated through years of theoretical and experimental research on topics relevant to scramjet propulsion. A previous volume with a similar organization, focused primarily introducing upper-level engineering students to the topic of hypersonic propulsion, was written by W. H. Heiser and D. T. Pratt more than a decade ago (*Hypersonic Airbreathing Propulsion*, AIAA Educational Series, J. S. Przemieniecki, editor, 1994). Considerable progress has been made in the intervening period. This book attempts to incorporate up-to-date advances made to understand the fundamental processes governing high-speed reacting flows and presents technological developments relevant to the scramjet engine.

Current developmental programs are briefly mentioned in the first chapter only to provide a general background of existing technological activities. The focus is on the phenomena that dictate the thermoaerodynamic processes encountered in the scramjet engine, including component analyses and flow-path considerations. Hence this book begins with theoretical background information pertaining to internal flow with chemical reactions and nonequilibrium effects, high-temperature gas dynamics, and hypersonic effects; trajectory, loads, and performance analyses are then reviewed, followed by cycle analyses. No single-engine cycle exists that can efficiently cover the whole range of a flight from takeoff to orbit insertion; therefore combined cycles are of particular interest for the design of the scramjet cycle. They are capable of providing the synergy required for increasing the efficiency of any individual propulsion cycle; therefore some of the more promising combined cycles are reviewed in the context of engine cycle analysis. Component analyses are further described, including inlets, nozzles, and isolators. The emphasis is then placed on the processes encountered in the combustion chamber. Current knowledge of injection, mixing, and mixing–combustion interactions is described, including some advanced modes of mixing enhancement, such as fuel preinjection upstream of the combustion chamber and reaction mechanisms for hydrogen and hydrocarbon compounds, including current reduced mechanisms for higher-order hydrocarbon fuels. Special attention is given to the structure of the recirculation region and the problem of flameholding, which is one of the key elements in the development of a feasible scramjet engine. A review of ground-based testing facilities, their capabilities, and applicability to the experimental study of scramjet engines follows. The considerable levels of energy associated with the operation of scramjet engines led to considerable difficulty in reproducing all the thermodynamic conditions encountered in flight in ground-based facilities. Theoretical modeling of physical processes, including the treatment of unsteady and high-temperature aerodynamics, has made substantial progress in the past decade and resulted in powerful computational fluid dynamic facilities. Finally, reviews of the current theoretical capabilities and issues are covered.

This book has benefited from contributions made by many individuals. There are many research collaborators to whom I am in particular debt for our shared work. At Pratt &Whitney-Rocketdyne, Allen Goldman, Atul Mathur, and Paul Ortwerth have been constant collaborators during my research. We had numerous conversations, and they have, in many instances, clarified observations related to flameholding and heat-release interactions in our common studies. Munir Sindir has a particular ability to distinguish the most relevant issues and has made, over the years, many useful research suggestions. Aaron Auslander from NASA Langley Research Center and I had long, edifying conversations related to hypersonic inlets and isolators. I am greatly indebted to Gabriel Roy, who, as a friend and program manager, has guided my and others' research for more than two decades. His vision and sense of scientific and technological relevance has led to the advancement of many propulsion engineering topics later emulated by researchers at national laboratories, universities, and industrial institutions. Viatcheslav Vinogradov from the Central Institute for Aviation Motors (CIAM) in Moscow, Russia, and I collaborated on several studies of hypersonic inlets with fuel preinjection, studies of mixing enhancement using thin pylons, and supersonic combustion studies with condensed- and gaseous-phase compounds. Joaquin Castro has made, on many occasions, relevant suggestions regarding flameholding of gaseous and condensed phases. Particular recognition should be given to the graduate students with whom I had the privilege of sharing the research at the University of Florida. They are a constant source of inspiration, and they keep reminding us, their advisors, of the purpose for our activities. This book has benefited greatly from contributions of Peter Gordon, a senior editor at Cambridge University Press. His patient editorial comments have vastly improved this manuscript.

I am indebted to Wei Shyy for the friendship and encouragement he has constantly shown me, particularly in the endeavor of writing this book. His work ethics and professional drive continue to be an inspiration to all of us who had the privilege of being his colleagues. Above all, I owe my gratitude to my wife, Anca, for her love and for the inspiration, the drive, the reason, the comfort, and the joy she has always been for me.

List of Acronyms

AFRL/PRA	U.S. Air Force Research Laboratory/Aerospace Propulsion Office
AIM	aerothermodynamic integration model
APL	Applied Physics Laboratory
ATREX	Air Turbo Ramjet Expander Cycle
CCP	combined-cycle propulsion
CDE	concept demonstration engine
CIAM	Central Institute for Aviation Motors
CPS	combination propulsion system
CUBRC	CALSPAN-UB Research Center
DAB	diffusion and afterburning
DCSCTF	Direct-Connect Supersonic Combustion Test Facility (NASA)
DCTJ	deep-cooled turbojet
DNS	direct numerical simulation
ESOPE	Etude de Statoréacteur comme Organe de Propulseur Evolué (France)
ETO	Earth-to-orbit
FMS	force measurement system
GALCIT	Graduate Aeronautical Laboratories at the California Institute of Technology
GASL	General Applied Science Laboratory
Hifire	Hypersonic International Flight Research Experiment
HOTOL	horizontal takeoff and landing
HRE	Hypersonic Research Engine (NASA)
HTF	Hypersonic Tunnel Facility (NASA)
IFTV	incremental flight test vehicle
JAXA	Japan Aerospace Exploration Agency
JHU	Johns Hopkins University
Kholod	hypersonic flying laboratory (Russia)

LACE	liquid–air collection engine or liquid–air-cycle engine
LACRRE	liquid–air collection rocket–ramjet engine
LEA	flight experimental vehicle (Russia)
LEM	linear eddy-mixing
LEO	low Earth orbit
LES	large-eddy simulation
MCH	methylcyclohexane
MHD	magnetohydrodynamic
NAL	National Aerospace Laboratories (Japan)
NASP	National Aerospace Plane
NIST-JANNAF	National Institute of Standards and Technology/Joint Army–Navy–NASA–Air Force
NRC	National Research Council of Canada
ONERA	Office National d'Études et de Recherches Aérospatiales (France)
PDF	probability density function
PLIFF	planar laser iodine-induced fluorescence
PR	pressure recovery
PREPHA	Program of Research and Technology for Airbreathing Hypersonic Aircraft (France)
PRSV	Peng–Robinson–Stryjek–Vera
RANS	Reynolds-averaged Navier–Stokes
RBCC	rocket-based combined cycle
SAM	structural assembly model
SCRAM	Supersonic Combustion Ramjet Missile
Shefex	Sharp Edge Flight Experiment
SMC	simultaneous mixing and combustion
SPARK	spectral analysis for reaction kinetics
SSTO	single-stage-to-orbit
TBCC	turbine-based combined cycle
TsAGI	Central Aerodynamic Institute (Russia)
TSTO	two-stage-to-orbit
USV-X	unmanned space vehicle–experimental
UTRC	United Technology Research Center
VULCAN	viscous upwind algorithm for complex flow analysis

1 Introduction

1.1 The Ramjet and the Supersonic Combustion Ramjet (Scramjet) Engine Cycle

An invention attributed to René Lorin of France in 1913 (Hallion, 1995), the ramjet is a remarkable air-breathing engine in its conceptual simplicity. Lacking moving parts and achieving air compression only through internal geometry change, it is capable of extending the operation beyond flight speed when the gas-turbine engine becomes inefficient. The ramjet does not, however, operate from takeoff, and its performance is low at subsonic speeds because the air dynamic pressure is not sufficient to raise the cycle pressure to the efficient operational values.

Above a flight speed of around Mach 3, cycles using rotating machinery, i.e., compressors, are no longer needed to increase the pressure, which can now be achieved by changes in area within the inlet and the diffuser leading to the combustion chamber. Engines without core rotating machinery can operate with a higher maximum cycle temperature as the limit imposed by the turbine presence on the cycle maximum temperature can now be increased. The ramjet cycle with subsonic air speed at the combustion chamber entrance becomes more efficient. As the speed further increases, the terminal shock associated with subsonic combustion leads to both significant pressure losses and elevated temperatures that preclude, in great part, recombination-reaction completion, thereby resulting in considerable energy loss. It becomes more efficient to maintain the flow at supersonic speed throughout the engine and to add heat through combustion at supersonic speed. Figure 1.1 shows the estimated specific impulse for several cycles as the flight Mach number increases (McClinton, 2002). The rocket-cycle specific impulse is included for comparison. The ramjet or the scramjet must be combined with another propulsion system for takeoff.

Schematically, the differences between subsonic and supersonic combustion ramjet engines are shown in Fig. 1.2. The subsonic conditions in the

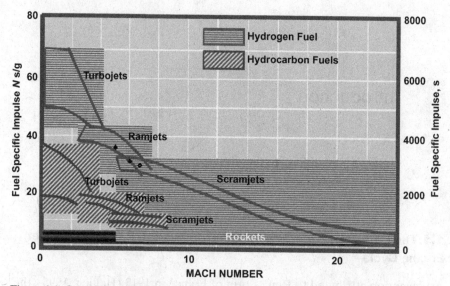

Figure 1.1. Specific impulses of several air-breathing cycles and rocket propulsion indicate the advantage of the scramjet engine over the other cycles for flight in excess of Mach 6. The diagram includes operation with hydrocarbon and hydrogen fuels.

combustion chamber in the former require the presence of a physical throat in the nozzle to maintain the desired inlet operational conditions, whereas the supersonic combustion chamber, in fact, requires an area increase as heat is released through combustion. For comparison, Table 1.1, offered by Ferri (1973), shows several critical parameters for the cases of supersonic vs. subsonic combustion at a selected flight condition: Mach 12 at an altitude of 40 km with hydrogen used as fuel, assumed to be in stoichiometric ratio with the engine airflow. The differences indicated in the table point to significant differences. The stagnation pressure recovery, which is a measure of the losses in the inlet and diffuser system, is about 30 times larger in the scramjet in comparison with the subsonic combustion ramjet because of the absence of the terminal normal shock. Because, in a first approximation, the engine thrust loses 1% for each 1% of loss in pressure recovery, the performance for the supersonic-combustion-based cycle is clearly evident. The temperature at the subsonic combustion chamber entrance is quite large. Severe dissociation is present at this temperature, and recombination reactions cannot take place within the combustion chamber. The net effect is, in fact, a reduction in

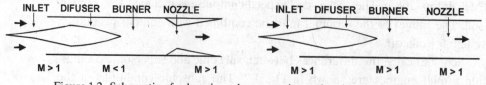

Figure 1.2. Schematic of subsonic and supersonic combustion ramjet engines.

Table 1.1. *A comparison of several relevant parameters between subsonic and supersonic combustion-based ramjets during Mach 12 flight*

Combustor chamber entrance	Supersonic	Subsonic	Combustion chamber exit	Supersonic	Subsonic
Ratio of burner entrance to capture area	0.023	0.023	Ratio of exit area to capture area	0.061	0.024
Stagnation-pressure recovery	0.5	0.013	Ratio of nozzle throat to capture area	0.061	0.015
Pressure (atm)	2.7	75	Pressure (atm)	2.7	75
Temperature (K)	1250	4500	Temperature (K)	2650	4200
Mach number	4.9	0.33	Mach number	3.3	0.38

temperature. Heat released because of fuel–air chemical reactions would occur in this case only further downstream in the nozzle, where, because of expansion, the temperature will decrease. Achieving chemical equilibrium within the nozzle so that the recovered heat can be converted into kinetic energy would require prohibitively long and therefore heavy nozzles. Thrust is further increased based on the ratio of the nozzle throat to capture area, which limits the amount of airflow through the engine in the subsonic-combustion-based ramjet. The scramjet will, in fact, substitute the mechanical throat with a thermal throat that results when the flow is slowed through tailored heat release. Finally, the considerably lower static pressure in the scramjet engine reduces the structural load on the engine duct, resulting in a lighter construction and overall increased system efficiency.

Technologically, the scramjet engine presents considerable difficulties that derive both from its operational characteristics and from the point of view of integration with the vehicle. Some of them are subsequently listed.

With air residence time of the order of milliseconds between engine capture and exit through the nozzle, fuel mixing time at the molecular level becomes a limiting factor. Mechanisms that accelerate mixing result in increased momentum losses, and they have to be traded for overall efficiency. The problem is compounded when liquid fuels are used because additional processes, including liquid breakup and vaporization, are present.

Flame stability becomes a key issue at high speeds and some kind of flameholder must be present when the residence time is increased. The chemical composition in the flameholding region is not only vastly different from the rest of the engine but is also characterized by large gradients in composition and temperature. Fuel–air-ratio tailoring must be such that the flameholding regions are stable for the entire range of the flight regime and engine-throttling conditions.

Prolonged operation as the vehicle accelerates through the atmosphere requires cooling of both the vehicle and engine components. The fuel on board will be the most appropriate candidate for this process to eliminate the need for a separate cooling agent and heat exchangers that would add to the vehicle's structural weight. In general, the engine fuel flow will not match exactly the cooling requirements, and some kind of fuel bypass will be required. Furthermore, for certain conditions, the fuel will not have the cooling capacity to satisfy the mission requirements: Heiser and Pratt (1994) indicate that, beyond Mach 10, hydrocarbon fuels can no longer satisfy the vehicle cooling requirements and cryogenic hydrogen would become, in this case, the fuel of choice.

Because neither subsonic nor supersonic combustion ramjets can operate from takeoff and produce competitive propulsive performances at low speeds, other thermodynamic cycles will be needed, either turbojets or rockets. If the mission includes operation beyond the Earth's atmosphere, rockets will have to be present on board. It is inefficient to incorporate several separate propulsive systems that operate in a certain sequence. Furthermore, various propulsive systems may be designed to operate in combined cycles, thereby achieving a synergetic enhancement of each individual cycle performance. Beyond the scramjet incorporation in combined-cycle architectures, the narrow shock-wave angles experienced during hypersonic flight make the entire vehicle forebody part of the engine intake system. The nozzle has a considerable length and will be part of the vehicle afterbody. There is thus a close interaction between the engine and the vehicle with the vehicle geometry and flight attitude that influences the engine airflow thermodynamic and flow-field conditions and the engine operation, in turn affecting the aerodynamic forces and moments experienced by the vehicle. The engine and the vehicle designs cannot therefore be uncoupled.

1.2 Historical Overview

The first design of an operational ramjet-engine-equipped airplane is René Leduc's demonstrator shown in Fig. 1.3, which was designed to separate from the airplane that brought it to altitude. Conceptually the design began in the 1920s, was patented in 1934 (Hallion, 1995), and immediately attracted the attention of the French government. World War II delayed its flight until 1946, and free-gliding tests began achieving powered climbs in 1949. At the time Leduc was developing his concept and actively pursuing the realization of his ramjet-equipped airplane, developmental work was taking place in the USSR, England, Germany, and the United States. Recognizing that the ramjet cycle becomes more efficient at higher speeds than the airplanes were capable of achieving at the time, experiments used projectile-launched

Figure 1.3. Leduc 0.10 at Musee de l'Air et de l'Espace at Le Bourget, France. The aircraft used a ramjet engine, and it was mounted on top of Languedoc aircraft for launch in flight.

ramjets or two-stage devices with a rocket-booster first stage separating from the ramjet-powered stage at high speeds. Liquid-fueled ramjet engines were used in the USSR in 1940 to boost the performance of a propeller-driven, Polikarpov I-152 biplane (Hallion, 1995; Sabel'nikov and Penzin, 2000), thus preceding Leduc's ramjet-powered flight to claim the first flight using ramjet-powered airplanes. Theoretical studies and experimentally projectile-launched ramjets were underway in Germany with Lippisch and Sänger's work in the 1940s and eventually reached Mach 4.2 at the end of the burnout (Avery, 1955). In the United States the early work of Roy Marquardt led to ramjet engines mounted on the wingtips of North American P-51 Mustangs as early as 1945, and later, larger versions of Marquardt ramjets were installed on the Lockheed P-80, allowing the airplane to fly under the ramjet power alone. Most of the ramjets developed in the following period focused on missile technology.

Along with the development of ramjet engines for airplane or missile applications, the concept of heat addition to a supersonic airstream took shape in the latter part of the 1940s. Captivating and presenting a valuable history of the scramjet development are Avery's article (1955), Hallion's report (1995), which covers the early scramjet research period through the Hypersonic Research Engine (HRE) program in the 1960s, and the articles by Waltrup et al. (1976) and Curran (2001), which describe scramjet-related activities in Australia, France, Germany, Japan, and the USSR. All these documents include ample references, including additional review publications.

In a 1958 study, which has become a point of reference, Weber and McKay noted that combustion can take place in supersonic airflows without creating considerable losses through shock-wave generation. Their study indicated that both the conventional ramjet and the scramjet efficiencies increase with speed in the range of Mach 4–7 and that the scramjet is more efficient than the ramjet above Mach 7; with an appropriately designed inlet, the scramjet advantage over the ramjet could be extended to Mach 5 flight. The results of this study identified many of the pertinent technical issues in the high-speed range, including the difficulties associated with flameholding in a supersonic flow, achieving an acceptable degree of mixing without causing severe shock losses, the significance of the inlet design on the cycle efficiency, the need to delay choking through heat release and thereby adopting a diverging combustion area, structural heat loads, and nozzle efficiency.

The work of Antonio Ferri at the beginning of the 1960s (Ferri, 1964) made a substantial contribution to the understanding of mixing and diffusive combustion processes in supersonic flows and was, to a large extent, the major driver for the technological developments that were about to arise. Ferri expanded on his earlier research in his review in a 1973 article indicating that, because the local temperature in the flame region is high, chemical reactions are fast compared with diffusion and heat conduction is due to mixing; therefore the process is dominated by transport properties. Although chemical kinetic rates are fast, the process nevertheless occurs at a finite rate and the reaction is distributed over an entire region in the flow; in regions of low pressure and temperature, the mixing and chemical time may become comparable and considerable mixing may take place before chemical reactions are completed, resulting in flame distribution over a large reaction zone. This heat release affects the pressure in the neighboring region and may even generate shocks in the unburned gas. Further, Ferri indicates that heat addition to a supersonic flow within a fixed geometry can be achieved efficiently for a broad range of flight Mach numbers, because the flow is less driven toward choking than it would be in the case of subsonic combustion; a three-dimensional design is thus capable of producing thrust efficiently if the geometry is chosen to correspond to the compression produced by combustion and, at the same time, satisfies the requirement for locally generating low Mach numbers for flame stability without substantial inlet contraction. The basis of the modeling of the physical processes is explained, emphasizing the three-dimensional nature of the flow field wherein finite-rate chemistry is coupled with the fluid dynamic processes that are dominated by the transport properties.

Large research projects were initiated in the early 1960s, most notably NASA's HRE Project (Andrews and Mackley, 1994). The goals were to build and test in flight a hypersonic research ramjet–scramjet engine using the X-15A-2 research airplane that was modified to carry hydrogen as the fuel for the scramjet engine. Two models were fabricated to test the structural engine integrity and to demonstrate the aerothermodynamic performance. The 8-ft., Mach 7 wind tunnel at NASA's Langley Research Center was used to test the structural assembly model (SAM), and the performance was evaluated on the aerothermodynamic integration model (AIM) at Mach 5–7 conditions at NASA's Glenn Research Center at the Plumbrook Hypersonic Test Facility. A pretest model is shown in Fig. 1.4. The SAM was evaluated with flightworthy hardware, and hydrogen was used as a coolant. Local heat and mechanical

Figure 1.4. A pretest model of the HRE.

loads were estimated, and material fatigue damage was assessed in more than 50 cycle loads. Thermal stresses expected in flight were duplicated in the wind tunnel, and a considerable database of surface temperature, cooling loads, and thermal fatigue was generated during this program. The AIM was a water-cooled, ground-based model with full simulation of Mach 5 and 6 enthalpy and reduced Mach 7 temperature. Both ramjet and scramjet operational modes were investigated, and critical technological areas were evaluated, including inlet boundary-layer transition, transition from subsonic to supersonic combustion and fuel distribution, and interactions between inlet and combustor and combustor and nozzle during transient operation.

The history of scramjet research in the USSR is just as old as that in the United States. Beginning with the work of Shchetinkov in the late 1950s (Sabel'nikov and Penzin, 2000) and continuing in the following decade, the Soviet researchers focused on the major issues encountered in the scramjet engine: chemical conversion efficiency at high temperatures, heat transfer at low-pressure conditions, and design operation efficiency. Shchetinkov and his group of researchers proposed using porous walls for fuel injection as a means both to address wall cooling and to reduce friction.

During this early research in the USSR, Shchetinkov's work identified supersonic combustion as dominated by mixing as the limiting factor and formulated solutions for the mixing length and the requirement for a divergent section to maintain a high level of efficiency. Furthemore, the contributions emerging from this group extended to analyses of combined cycles that included scramjet operation, including ram-rockets and atmospheric air collection (later known as liquid–air collection engines – LACEs).

At the time when NASA was studying the HRE concept, a joint NASA/U.S. Air Force working group recognized the potential of the scramjet technology and set a common goal to pursue the technology that would result in a scramjet-operated vehicle within the 1960s (Hallion, 1995). This program was based both on the research evolving around the HRE program and on flight testing of hypersonic engine and airframe models by use of the X-15 airplane. These plans were severely impaired by the cancellation of the X-15 programs and came to an end toward the latter part of the decade. During the same period, however, the U.S. Air Force evaluated several scramjet engines in ground testing, including a variable-geometry Mach 5 engine developed by UTRC, a Mach 7 component integration engine developed by Ferri at GASL, and a Marquardt flight-weight dual-mode combustion scramjet (Waltrup et al., 1976). The Scramjet Incremental Flight Test Vehicle (IFTV) (Peschke, 1995), although tested in flight in only an unpowered configuration, produced valuable advances concerning fuel-injection tailoring for engine heat release and inlet compatibility during component ground testing. Hallion (1995) describes in detail the ground developmental testing and the aerodynamic nonpowered flights accomplished during this exciting program, which took the concept to flight hardware.

The axisymmetric configuration was also evaluated by Soviet (Vinogradov et al., 1990) and French researchers during the ESOPE program and was used in later international flight-testing programs (Voland et al., 1999; Falempin, 2000). Several flight tests of axisymmetric scramjet models boosted by SA-5 rockets took place in 1991 and 1992 in collaboration with ONERA researchers and were repeated in 1998 (Voland et al., 1999) as part of the Central Institute for Aviation Motors (CIAM) in Moscow and NASA interactions. All these engines were based on cavities for flameholding and used distributed hydrogen fuel injection to optimize the axial heat release. The original configuration used in CIAM studies is described in detail by Vinogradov et al. (1990). The ESOPE program in France, an axisymmetric hydrogen-fueled scramjet in the early 1970s focused, as in the United States and the USSR, on mixing-efficiency improvements and, similar to the U.S. HRE Program, ended following ground testing before flight-test hardware was built.

The airframe-integrated scramjet concept that emerged in subsequent years led to NASA's rectangular scramjet configuration. This configuration generated a complex inlet-flow structure and included in-stream struts with fuel injectors that could modulate the heat addition as required by the flight regime. This concept, which was evaluated extensively at NASA during the 1970s (Northam and Anderson, 1986), was later adopted in other programs [for example, National Aerospace Laboratories (NAL) studies in Japan; see Chinzei et al., 2000]. Figure 1.5 shows the rectangular engine configuration that is suitable for modular engine design and is particularly attractive for integration with the airframe when the application is in a transatmospheric vehicle. The swept inlet cowl provides flow stability over a large flight regime, and fuel-injection modulation from the struts allows operation over a broad Mach number range with a fixed geometry.

Figure 1.5. The rectangular engine configuration used for the National Aerospace Plane Program's concept demonstration engine mounted atop its pedestal and the six-component force measurement system in preparation for testing in NASA Langley's 8-ft. high-temperature tunnel.

If the rectangular-shaped engine is appropriate for integration in a larger airframe, for a transatmospheric vehicle, the axisymmetric engine is well suited for a hypersonic missile. The Supersonic Combustion Ramjet Missile (SCRAM) Program developed in the latter part of the 1960s and early 1970s at JHU/APL (Silver Spring, MD) used a compact design with the scramjet engine surrounding the missile components. A contoured inlet was designed to provide starting and stability over the entire flight regime and the internal area distribution included an isolator to protect the inlet flow from the pressure rise in the combustion chamber (Billig, 1993). This program made a substantial contribution to the study of shock-train pressure rise and the interactions between shock waves and boundary layers in the isolator. Combustion modeling studies performed during this program played an important role in establishing design criteria for supersonic combustion chambers.

The emergence of the National Aerospace Plane (NASP) Program in the United States gave a new effervescence to hypersonic activities. The concept of a single-stage-to-orbit (SSTO) vehicle using air-breathing propulsion for the transatmospheric part of the trajectory, with rocket propulsion for the final insertion into orbit, was an extension of earlier concepts of rocket-based, entirely reusable SSTO concepts at Boeing and Rockwell (Hallion, 1995) with the addition of air-breathing propulsion. Figure 1.6 shows a concept of a proposed NASP configuration as anticipated toward the end of the 1980s. The NASP represented a significant step forward from the Space Shuttle: Using horizontal takeoff and landing, its operation resembled that of an airplane more than that of a rocket; fully reusable for more than 150 flights, it was designed to operate efficiently both during ascent and during maneuvering at

Figure 1.6. An artist's rendition of the NASP.

high altitudes, making maneuvers and orbital changes; finally, it would reenter and land under its own power.

The NASP posed technological challenges in materials, propulsion, aero-dynamics, sustainability, and flight control, and, as a result, it became a cat-alyst for noticeable advances made in all these areas. New composite mate-rials were developed to satisfy requirements for lightweight and structural resistance and to optimize their performance by minimizing weight as well as increasing load-carrying capacity. Metallic and carbon foams were con-ceived to produce materials with a wide range of thermal conductivities for use in thermal protection systems as well as in heat exchangers, which are nec-essary during extended hypersonic flights to maintain vehicle integrity. New concepts of combined-cycle propulsion systems evolved such that synergis-tic advantages can be extracted, and extensive testing was undertaken in the Mach 4–7 range. Air-breathing-propulsion-related high-Mach-number exper-iments were undertaken in shock and expansion tunnels. But perhaps the most significant achievement of the research undertaken during this project was the development of predictive tools in the area of computational fluid dynamics, with applications to both external aerodynamics and internal flows with chemical-reaction modeling for propulsion applications. Supported by an unprecedented development of computing power, predictive models for both fluid and solid mechanics have advanced, including both numerical schemes and the modeling of the physical processes. Today the degree of accuracy acquired by these models allows their integration into the early stages of the design process.

The PREPHA program in France emerged at about the time the NASP Program in the United States was approaching its end. With the vision toward the SSTO vehicle, the program has made advances in several areas, including material testing, modeling, and development of ground-testing capabilities (Marguet et al., 1997). Conceptual analyses of combined cycles included turborocket – scramjet and dual-mode rocket – scramjet combined cycles. Substantial system studies undertaken during this program identified the rocket-based combined-cycle (RBCC) system as the more efficient propulsion cycle for a transatmospheric vehicle. Experimental studies focused on mixing and combustion using hydrogen at Mach 6.5 flight conditions. The results contributed to the improvement of the physical models used in theoretical analyses. Component studies included shock–boundary-layer, isolator–combustion chamber, and nozzle–afterbody interactions. Several systems have been evaluated during successive programs (Lentsch et al., 2003), and combined cycles that most appropriately respond to the mission requirements have been analyzed.

In Germany, the SÄNGER Program initiated in 1987 envisioned a two-stage-to-orbit (TSTO), fully reusable concept, using a TSTO reusable vehicle (Bissinger et al., 2000). In addition to the system and architecture analyses, this program was a vehicle for the development of computational models and included an interaction with the Russian researchers from TsAGI to evaluate isolator–combustion chamber interactions under a range of injection configurations.

Substantial evaluation of supersonic combustion at high enthalpy has been contributed by the research performed in Australia under the leadership of R. J. Stalker in the T4 shock tunnel. With capabilities of simulating orbital flight enthalpy, this research covered fuel mixing and combustion, fundamental skin friction, strong shock–boundary-layer interactions, effects of shock waves on mixing and combustion, and thrust generation. Along with hydrogen, hydrocarbon fuels were evaluated, and effects of hypergolic additives were investigated (Paul and Stalker, 1998). The flight experiment using the axisymmetric HyShot configuration in 2003 produced data at a flight Mach number of ∼7.5 for comparison with similar information obtained on ground facilities (Boyce et al., 2003). The vehicle, boosted by a two-stage rocket, was accelerated beyond the atmosphere and redirected toward the ground. The scramjet, which included a boundary-layer bleed on the intake and a constant-area combustor, operated after reentry for approximately 5 s between 35 and 23 km.

Japan has had activities related to supersonic combustion since the late 1970s, including evaluations of flame development under various injection configurations. The 1990s saw a significant investment in facilities for ground testing, including the development of a Mach 8–15 high-enthalpy shock

Figure 1.7. The first flight-weight hydrocarbon-fueled scramjet in the GASL wind tunnel was tested at Mach 6.5 with JP-7 used to cool the engine components prior to injection into the combustion chamber.

tunnel. A rich activity in all relevant areas including engine components, system integration, operability, and fundamental research followed (Chinzei et al., 2000). Considerable work was done in the NAL, later renamed the Japan Aerospace Exploration Agency (JAXA), using a rectangular engine configuration with vertical struts reminiscent of NASA's configuration evaluated during the NASP Program.

In the wake of NASP Program cancellation in the United States several projects followed, mostly focused on the lower hypersonic operational range. Among them, the U.S. Air Force HyTech program resulted in testing at Mach numbers up to 6.5 of a hydrocarbon-fueled scramjet with flight-weight hardware, including fuel-cooled combustor components. Figure 1.7 shows the GRE-1 engine in the wind tunnel at GASL.

The successful flight in March 2004 of the X-43 represented the first successful attempt to fly a configuration that included an integrated hypersonic airframe–scramjet engine, and it is likely to provide invaluable flight data that are otherwise impossible to acquire in ground-testing facilities.

1.3 Summary

The progress made since the early days of space exploration resulted in the knowledge for building rocket-based vehicles in a range of sizes capable of responding to the various missions' requirements, ranging from placing constellations of small satellites on an Earth orbit to interplanetary flight. Common to all these launches is the low degree of recovery or reusability of major parts. Reusable in large part, the U.S. Space Shuttle and the Russian Energiya/Buran vehicles were designed decades ago, and these systems' operability and maintenance costs are large, which translate into a high price per pound of payload launched into orbit. Lowering these costs is a major factor in the current philosophy for the design of future vehicle configurations.

The next generation of launch vehicles will be required to be entirely reusable, capable of maneuvering at all altitudes and flight speeds and easily prepared for the next launch. In this category, transatmospheric vehicles that include air-breathing propulsion for at least part of their mission appear particularly attractive. The ability to fly using atmospheric lift and an ambient oxidizer leads to an increase of the payload weight per total takeoff gross by a factor of two when air-breathing architectures are compared with all rocket-based vehicles (Czysz and Vanderkerckhove, 2000).

A large part of the transatmospheric acceleration must be supported by a ramjet cycle with supersonic flow throughout the engine – a scramjet. The other stages of the mission, including takeoff, acceleration, and orbital insertion, will be based on combined cycles. As Czysz and Vanderkerckhove (2000) point out, on one hand, the technology appears to be within reach if operational affordability and reliability can be solved. On the other hand, the physical processes that take place in the scramjet engine are more complex than in most of the other propulsion systems. The short residence time, the large velocity gradients that result in subsonic regions embedded in supersonic flows, with a closely coupled interaction between mixing and chemical kinetics, with strong shock wave–boundary-layer interactions still must be solved in a satisfactory manner for the entire operational range.

Recent years have witnessed several flight tests that each, in turn, expanded the knowledge and capability. The GELA flight test of a maneuvering scramjet missile in Russia in 2004, which followed the earlier test in 2001, the successful flight test of the HyShot program in 2003, and the remarkable flight test of the integrated vehicle–scramjet experimental X-43 in the United States in 2004 indicate that scramjet technology has matured to a degree that gives increased confidence in the design capabilities for flight. In the early days of exploration beyond Earth's boundaries, the imagination and national drive could be lifted by the vision of great human achievement.

Several international programs are currently advancing with promising results. The U.S.–Australian collaboration for the Hypersonic International Flight Research Experiment (Hifire) follows the HyShot program with the goal of gathering flight data at reentry speed conditions. In France, LEA is a collaborative effort between MBDA and ONERA with Russian scientists to develop a vehicle that will fly in the range of Mach 4–8 with future expansion to the Mach 10–12 regime. In the United States, the X-51A hydrocarbon-fueled engine has undergone several ground-testing programs with flightworthy weight components and is being prepared for flight tests in the near future; it is expected to sustain as much as 4 min of scramjet-powered operation.

To reach suborbital altitudes, vehicles capable of attaining velocities of several kilometers per second need to be developed. Activities in this regime also exist. The USV-X program in Italy is one example; Germany's Sharp Edge Flight Experiment (Shefex) demonstrates materials' capability of withstanding thermal stresses for resusable vehicles in the Mach 10–11 regime (Norris, 2008).

Today it has become clear that there is a need to develop a next generation of reusable vehicles that reach orbit at substantially lower cost and increased reliability; flight through the atmosphere at more then twice the currently possible speed has also been identified for defense applications. Heppenheimer (1999) quoted George E. Mueller, a former associate administrator for Manned Space Flight, who said in 1969 in his review of the process leading to the development of the Space Shuttle that NASA had set for itself the goal to reduce the then $1,000 per pound of payload delivered in orbit to a level of less than $50 a pound. With adjustments for inflation, that goal still stands. Technology has made significant progress in this interval. The decision-making process still has to catch up with it.

REFERENCES

Andrews, E. H. and Mackley, E. A. (1994). "NASA's Hypersonic Research Engine Project – A review," NASA TM-107759.

Avery, W. H. (1955). "Twenty years of ramjet development," *Jet Propul.* **25**, 604–614.

Billig, F. S. (1993). "Research on supersonic combustion," *J. Propul. Power* **9**, 499–514.

Bissinger, N. C., Koschel, W., Sacher, P. W., and Walther, R. (2000). "SCRAM-jet investigations within the German Hypersonic Technology Program (1993–1996)," in *Scramjet Propulsion* (E. T. Curran and S. N. B. Murthy, eds.), Vol. 189 of Progress in Astronautics and Aeronautics, AIAA, pp. 119–158.

Boyce, R., Gerard, S., and Paull, A. (2003). "The HyShot Scramjet Flight Experiment – Flight data and CFD calculations compared," AIAA-2003–7029.

Chinzei, N., Mitani, T., and Yatsuyanagi, N. (2000). "Scramjet engine research at the National Aerospace Laboratory in Japan," in *Scramjet Propulsion* (E. T. Curran and S. N. B. Murthy, eds.), Vol. 189 of Progress in Astronautics and Aeronautics, AIAA, pp. 159–222.

Curran, E. T. (2001). "Scramjet engines: The first forty years," *J. Propul. Power* **17**, 1138–1148.

Czysz, P. and Vanderkerckhove, J. (2000). "Transatmospheric launcher sizing," in *Scramjet Propulsion* (E. T. Curran and S. N. B. Murthy, eds.), Vol. 189 of Progress in Astronautics and Aeronautics, AIAA, pp. 979–1104.

Falempin, F. H. (2000). "Scramjet development in France," in *Scramjet Propulsion* (E. T. Curran and S. N. B. Murthy, eds.), Vol. 189 of Progress in Astronautics and Aeronautics, AIAA, pp. 47–117.

Ferri, A. (1964). "Review of problems in application of supersonic combustion," *J. R. Aeronaut. Soc.* **68**, 575–597.

Ferri, A. (1973). "Mixing controlled supersonic combustion," *Annu. Rev. Fluid Mech.* **5**, 307–338.

Hallion, R. P. (1995). *The Hypersonic Revolution, Volume II: From Max Valier to Project Prime*, Aeronautical System Center, Air Force Material Command, Wright Patterson Air Force Base, ASC-TR-95–5010.

Heiser, W. H. and Pratt, D. T. (with Daley, D. H. and Mehta, U. B.). (1994). *Hypersonic Airbreathing Propulsion*, AIAA Educational Series.

Heppenheimer, T. A. (1999). *The Space Shuttle Decision*, NASA History Series, NASA History Office, NASA SP-4221.

Lentsch, A., Bec, R., Deneu, F., Novelli, P., Rigault, M., and Conrardy, J. (2003). "Airbreathing launch vehicle activities in France – The last and the next 20 years," AIAA-2003–6949.

Marguet, R., Cazin, P., Falempin, F. H., and Petit, B. (1997). "Review and comment on ramjet research in France," in AIAA 97–7000, pp. 3–13.

McClinton, C. R. (2002). "Hypersonic technology...Past, present and future," a presentation at the Institute for Future Space Transport, Gainesville, FL. Available from C. Segal.

Northam, G. B. and Anderson, G. Y. (1986). "Supersonic combustion research at Langley," AIAA Paper 86–0159.

Norris, G. (2008, July 14). "International hypersonic research and development may be a key to achieving high-speed airbreathing flight," *Aviat. Week Space Technol.*, 128–132.

Paul, A. and Stalker, R. J. (1998). "Scramjet testing in the T4 Impulse Facility," AIAA Paper 98–1533.

Peschke, W. O. T. (1995). "Approach to in situ analysis of scramjet combustor behavior," *J. Propul. Power* **11**, 943–949.

Sabel'nikov, V. A. and Penzin, V. I. (2000). "Scramjet research and development in Russia," in *Scramjet Propulsion* (E. T. Curran and S. N. B. Murthy, eds.), Vol 189 of Progress in Astronautics and Aeronautics, AIAA, pp. 223–368.

Vinogradov, V. A., Grachev, V., Petrov, M., and Sheechman, J. (1990). "Experimental investigation of 2-D dual mode scramjet with hydrogen fuel at Mach 4...6," in *Proceedings of the AIAA Second International Aerospace Planes Conference*, AIAA.

Vinogradov, V. I. (1999). "Design analysis of the model hydrocarbon fueled engine performance," JHU/APL, Final Contractor Report No. 700–773723, New Technologies Implementation Center, Central Institute of Aviation Motors.

Voland, R. T., Auslender, A. H., Smart, M. K., Roudakov, A. S., Semenov, V. I., and Kopchenov, V. (1999). "CIAM/NASA Mach 6.5 scramjet flight and ground test," AIAA Paper 99–4848.

Waltrup, P. J., Stull, F. D., and Anderson, G. Y. (1976). "Supersonic combustion ramjet (scramjet) engine development in the United States," in *Proceedings of the Third International Symposium on Air Breathing Engines*, AIAA, pp. 835–862.

Weber, R. J. and McKay, J. S. (1958). "An analysis of ramjet engines using supersonic combustion," NACA TN-4386.

2 Theoretical Background

Because the processes in the scramjet engines are governed by a mixture of fluids in motion undergoing chemical reactions, the fluid mechanics conservation equations along with transport properties and chemical kinetics must be solved simultaneously. This chapter treats the governing equations with distinction between field equations, also called conservation laws and constitutive equations; the one-dimensional (1D) flow simplification, which is often used when some fundamental processes in a scramjet engine are described, is also included. Equilibrium chemistry and the departure from equilibrium follow.

2.1 Field Equations and Constitutive Relations for Compressible Flows

The equations of motion, along with the constitutive equations, provide a system of mathematical expressions that model physical fluid properties under certain given boundary conditions.

2.1.1 Field Equations of Fluid Motion

The field equations of motion, also called conservation laws, are derived for gases or liquids, which, for dynamics studies, are similarly treated under the fluids category. Both exhibit the property of easy deformability dictated by the nature of the intermolecular forces. A detailed discussion of the distinction between gases and fluids is included, for example, in Batchelor (1994) and is not reproduced here. The fundamental assumption used in the derivation is that of continuum, granted by the sufficiently high density of the fluid, which eliminates the distinction between individual molecules and the intermolecular space within the entire volume of fluid under consideration.

The field equations include conservation of mass, momentum – also referred to as the Navier–Stokes equations – and energy. The field equations apply to a volume of fluid with distinction of specific fluid properties. The description of these equations depends on surface equilibria and may be subject to surface rate processes. However, in most cases the physical processes are reduced to interface interactions, and simple expressions for mass, momentum, and energy conservation can be derived (Cullen, 1985). These equations are summarized in the next subsection.

2.1.1.1 Mass Conservation

In an elemental volume of the fluid enclosed by a surface, the instantaneous amount of mass is balanced by the amount of fluid crossing the surface. Mathematically, this balance can be written as

$$\frac{D}{Dt} \int_V \rho \, dV = 0, \tag{2.1}$$

where the operator $\frac{D}{Dt}$ represents the material derivative defined as

$$\frac{D}{Dt} = \frac{\partial}{\partial t} + \bar{u}\nabla = \frac{\partial}{\partial t} + u_k \frac{\partial}{\partial_k}. \tag{2.2}$$

If the material derivative is introduced into the volume integral, the mass conservation takes the form of

$$\int_v \left[\frac{\partial \rho}{\partial t} + \frac{\partial (\rho u_k)}{\partial x} \right] dV = 0. \tag{2.3}$$

Because the volume V is arbitrary, it follows that the integrand is zero everywhere; hence the continuity equation can be written as

$$\frac{\partial \rho}{\partial t} + \frac{\partial (\rho u_k)}{\partial x} = 0. \tag{2.4}$$

2.1.1.2 Momentum Conservation Equations

The momentum time chance is balanced by the sum of all forces acting on the surface that bounds an element of fluid and can be written as

$$\rho \frac{D\bar{u}}{Dt} = -\nabla p + \nabla \tilde{\tau} + \bar{F}_b. \tag{2.5}$$

As written, the momentum equations include the hydrostatic pressure acting on the volume ∇p, the viscous stress tensor that acts at the fluid element boundary $\nabla \tilde{\tau}$, and the body forces $\bar{F}_b$. It should be noted that, although the transport properties need to be calculated to determine the shear stress, the body forces must be defined for every case.

In tensor form, the momentum conservation can be written as

$$\rho \frac{Du_i}{Dt} = -\frac{\partial p}{\partial x_i} + \frac{\partial \tau_{ik}}{\partial x_k} + F_{b_i}. \tag{2.6}$$

2.1.1.3 Conservation of Energy

The conservation of energy requires that any change in energy be related to work done on the element of fluid and heat exchanged with the surroundings. The equation can be written as

$$\rho \frac{D}{Dt}\left(e + \frac{u^2}{2}\right) = -\nabla\left(p\bar{u}\right) + \nabla\left(\bar{\tau}\bar{u}\right) - \nabla\bar{q} + \bar{F}_b\bar{u} + Q, \tag{2.7}$$

where $\nabla\bar{q}$ represents the heat transferred through conduction and Q is the bulk heat addition. In tensor form, the energy conservation equation is written as

$$\rho \frac{D}{Dt}\left(e + \frac{u^2}{2}\right) = -\frac{\partial}{\partial x_i}\left(pu_i\right) + \frac{\partial}{\partial x_k}\left(\tau_{ik}u_i\right) - \frac{\partial q_i}{\partial x_i} + F_{b_k}u_k + Q. \tag{2.8}$$

When the enthalpy $h = e + pv$ is used, the conservation of energy becomes

$$\rho \frac{D}{Dt}\left(h + \frac{u^2}{2}\right) = \frac{\partial p}{\partial t} + \frac{\partial}{\partial x_k}\left(\tau_{ik} \cdot u_i\right) - \frac{\partial q_i}{\partial x_i} + F_{b_k} \cdot u_k + Q. \tag{2.9}$$

A particular case arises for steady, inviscid flows in the absence of heat transfer and body forces, reducing the energy equation to

$$\frac{D}{Dt}\left(h + \frac{u^2}{2}\right) = 0 \quad \text{or} \quad h + \frac{u^2}{2} = \text{const.}, \tag{2.10}$$

indicating that, in this case, the total enthalpy is constant along a streamline.

The conservation equations for a three-dimensional (3D) space include five equations and 15 unknowns. The constitutive equations provide additional equations to complete the system.

2.1.1.4 Conservation of Species

In multicomponent reacting gas mixtures the field equations need to be complemented with the equation of conservation of species:

$$\frac{\partial Y^i}{\partial t} + \bar{u}\nabla Y^i = \frac{w^i}{\rho}, \quad i = 1, \dots, N, \tag{2.11}$$

where the superscript i refers to species i and the diffusion velocities have been neglected. The production terms w^i derive from the chemical kinetics analyses described in Chap. 6.

2.1.2 Constitutive Equations

If the field equations apply to any fluid, the constitutive equations are specific to the particular fluid under study. Often the mixture of gases in the scramjet engines is treated as an ideal gas; the constitutive equations take the following form.

2.1.2.1 Equations of State

The equation of state provides a relation of thermodynamic variables in equations of the type

$$p = p(\rho, T, N_i). \tag{2.12}$$

For a thermally perfect gas, which assumes nonintermolecular forces and negligible molecule size, the equation of state simplifies to

$$p = \rho RT, \tag{2.13}$$

which assumes point-sized molecules and neglects the effect of the intermolecular forces on the pressure.

A state variable used in open systems to describe energy exchange, the enthalpy, is defined as

$$h = h(T, p). \tag{2.14}$$

At moderate temperatures, of the order of 1000 K, the fluid can be considered calorically perfect and the specific heat is considered temperature independent; therefore $h = c_p T$. The assumptions of thermally and calorically perfect gases represent the simplification known as ideal gases.

2.1.2.2 The Fourier Law for Heat Transfer

Conduction heat transfer can be described through the Fourier law of heat transfer, an empirical-based derivation that relates the heat transferred to the material properties and an applied temperature gradient,

$$q_i = -k \frac{\partial T}{\partial x_i}, \tag{2.15}$$

where the negative sign indicates that heat travels in the direction of decreasing temperature (Incropera and DeWitt, 2002).

2.1.2.3 The Shear-Stress Tensor

The shear-stress tensor derived for a Newtonian fluid reflects the proportionality between the shear stress and the rate of deformation:

$$\tau_{ik} = \mu \left(\frac{\partial u_i}{\partial x_k} + \frac{\partial u_k}{\partial x_i} \right) - \frac{2}{3} \mu \left(\nabla \bar{u} \right) \delta_{ik}. \tag{2.16}$$

The constitutive equations contribute 11 equations and introduce a single new variable, T, thereby completing, along with the field equations, a system of 16 equations with 16 unknowns. These equations are further simplified for particular fluid conditions. For inviscid flow $\mu \to 0$, the shear-stress tensor disappears on the right-hand side in momentum conservation equation (2.7). In many applications, body forces are neglected and heat transfer through conduction or bulk heat addition is not present, further simplifying the right-hand side of the momentum equation to include only the hydrostatic pressure.

2.2 One-Dimensional Steady Flow and the Rankine–Hugoniot Relations

2.2.1 One-Dimensional Steady Flow

The assumption of 1D flow indicates that all vectors, including velocity, heat flux, and body forces, are parallel to each other and the properties are uniform within surfaces perpendicular to this direction. The steady-flow assumption further indicates the absence of time dependence. Symbolically,

$$\frac{\partial}{\partial x_2} = \frac{\partial}{\partial x_3} = 0, \quad u_2 = u_3 = 0, \quad \text{1D assumption;}$$

$$\frac{\partial}{\partial t} = 0, \qquad \text{steady-flow assumption.}$$

Under these assumptions the field equations simplify as follows. Continuity becomes

$$\frac{d}{dx}(\rho u) = 0; \tag{2.17}$$

momentum conservation is

$$\rho u \frac{du}{dx} = -\frac{dp}{dx} + \frac{d\tau_{11}}{dx} + F_b; \tag{2.18}$$

and energy conservation is

$$\rho u \frac{dJ}{dx} = \frac{d}{dx}(\tau_{11} \cdot u) - \frac{dq}{dx} + F_b \cdot u + Q. \tag{2.19}$$

When a 1D steady flow experiences a change within a given domain that separates uniform regions, the equations can be integrated across the domain, thus relating the properties in the second region to the initial conditions. The transformations may include changes that are due to shock waves or chemical transformations and involve heat release, conduction, diffusion, and viscous effects. Because the transformations occur, in general, over a short distance, the fluid properties in uniform regions 1 and 2 in Fig. 2.1 are assumed inviscid and non-heat-conducting. The diagonal terms in the pressure tensor reduce to the hydrostatic pressure, and heat transfer through conduction or bulk addition is negligible.

These simplifications imply that

$$\left.\frac{d}{dx}\right|_1 = \left.\frac{d}{dx}\right|_2 = 0, \quad \left.\frac{d}{dx}\right|_{1 \to 2} \neq 0.$$

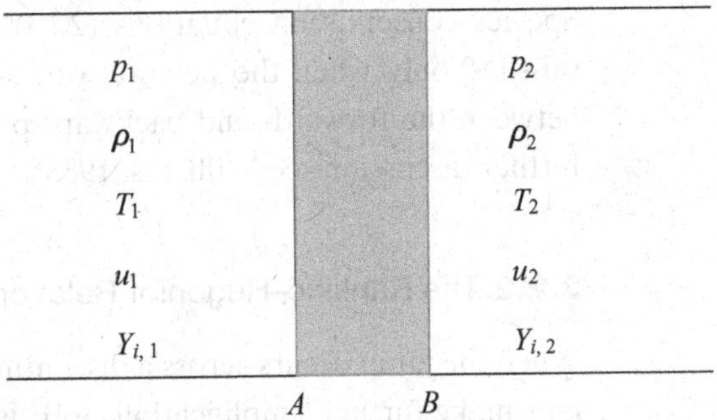

Figure 2.1. Schematic diagram of jump
conditions from a uniform region 1 to
a uniform region 2. Heat conduction,
viscous effects, chemical reactions, and
diffusion are present in the nonuniform
region $A \rightarrow B$.

p_1 $\qquad$ p_2

ρ_1 $\qquad$ ρ_2

T_1 $\qquad$ T_2

u_1 $\qquad$ u_2

$Y_{i,1}$ $\qquad$ $Y_{i,2}$

$A \qquad B$

Furthermore, inviscid flow and heat conduction are neglected,

$$\tau_{11} = 0,$$

$$q = 0$$

in regions 1 and 2.

The conservation equations can be integrated across the nonuniform
region and result in the following relations. For mass conservation,

$$\int_A^B \frac{d}{dx}(\rho u)\, dx = 0 \qquad (2.20)$$

or

$$\rho u|_A^B = 0.$$

For momentum conservation,

$$\int_A^B \frac{d}{dx}(p + \rho u^2 - \tau_{11})dx = \int_A^B F_b dx \qquad (2.21)$$

or

$$p + \rho u^2|_A^B = \int_A^B F_b.$$

Examples of body forces in the nonuniform region could be Lorentz forces,
diffusion produced by Dufour or Soret effects, or another type of force.

The energy conservation equation takes the form

$$\int_A^B \frac{d}{dx}(\rho u J - \tau_{11} u + q)\, dx = \int_A^B (F_b \cdot u + Q)dx \qquad (2.22)$$

or

$$\rho u J|_A^B = \int_A^B (F_b \cdot u + Q)dx.$$

For chemically reacting systems, the additional condition of species conservation is required:

$$w_1^i = w_2^i = 0, \quad i = 1, \dots, N. \qquad (2.23)$$

Species conservation equations (2.23) may not all be independent and are satisfied only when the net rate of reactions, if present, i.e., the difference between the forward- and backward-propagating reactions, is negligible (for further discussion see Williams, 1985).

2.2.2 The Rankine–Hugoniot Relations

When the jump occurs across a discontinuous region, such as a shock wave, we may make further simplifications to Eqs. (2.20)–(2.22) by assuming negligible body forces and the absence of bulk heat transfer. The conservation equations become (Liepmann and Roshko, 1957), after integration,

$$\rho_1 u_1 = \rho_2 u_2 \equiv m,$$
$$p_1 + \rho_1 u_1^2 = p_2 + \rho_2 u_2^2, \tag{2.24}$$
$$h_1 + \frac{u_1^2}{2} = h_2 + \frac{u_2^2}{2}.$$

These three equations include four unknowns, i.e., pressure, density, velocity, and enthalpy, in region 2. The equation of state and the enthalpy equation provide two additional equations with one additional unknown, the temperature, and thus form a determined set.

We can recast the jump conditions described by Eqs. (2.24) in terms of a ratio of similar gas properties across the shock, substituting velocity ratios for density ratios from the continuity equation to obtain

$$\frac{p_2}{p_1} = 1 + \frac{\rho_1 u_1^2}{p_1}\left(1 - \frac{\rho_1}{\rho_2}\right), \tag{2.25}$$
$$\frac{h_2}{h_1} = 1 + \frac{u_1^2}{2h_1}\left[1 - \left(\frac{\rho_1}{\rho_2}\right)^2\right].$$

For nonideal gases, we may use an iterative calculation to solve these equations, beginning with an estimate for the density ratio ρ_1/ρ_2; then we calculate the pressure p_2 and the enthalpy h_2 in the second region and recalculate the density across the discontinuity ρ_2 from the assumed equation of state. The calorically perfect-gas simplification assumes that the specific heat remains constant for the entire range of temperatures of interest and that the enthalpy in Eq. (2.14) can be written as $h = c_p T$. When both assumptions of thermally and calorically perfect gas are considered, the medium is called an ideal gas. In this case, we can express Eqs. (2.13) and (2.14) explicitly as functions of the Mach number in front of the shock by using the definition of the speed of sound:

$$M \equiv \frac{u}{a}, \quad a^2 = \left(\frac{\partial p}{\partial \rho}\right)_s. \tag{2.26}$$

Assuming that isentropic conditions exist on each side of the discontinuity, the speed of sound derived from the isentropic relation becomes

$$p = \text{const } \rho^\gamma \quad \Rightarrow \quad a^2 = \gamma \frac{p}{\rho} = \gamma RT. \tag{2.27}$$

After further manipulation, the jump conditions can be written as functions of the Mach number in front of the shock:

$$\frac{p_2}{p_1} = 1 + \frac{2\gamma}{\gamma + 1}\left(M_1^2 - 1\right),$$

$$\frac{T_2}{T_1} = \left[1 + \frac{2\gamma}{\gamma + 1}\left(M_1^2 - 1\right)\right]\frac{(\gamma - 1)M_1^2 + 2}{(\gamma - 1)M_1^2}, \tag{2.28}$$

$$\frac{\rho_2}{\rho_1} = \frac{(\gamma + 1)M_1^2}{(\gamma - 1)M_1^2 + 2}.$$

These equations indicate that, for very strong shocks as $M_1 \to \infty$, the density ratio is limited by

$$\frac{\rho_2}{\rho_1} = \frac{\gamma + 1}{\gamma - 1}, \tag{2.29}$$

resulting in the well-known limit (Liepmann and Roshko, 1957) of six for air considered nondissociated ($\gamma = 1.4$). At hypersonic speeds, the temperature rise across the shock is sufficiently large to cause dissociation and would then increase the density ratio given by Eq. (2.29). Considering the relatively small values for the density ratio in Eqs. (2.25) in comparison with the dynamic terms, including the velocity, it follows that most of the contribution to pressure and enthalpy is contained in the dynamic component and the following further approximations can be made,

$$p_2 \simeq \rho_1 u_1^2, \quad h_2 \simeq u_1^2/2, \tag{2.30}$$

which indicates that the pressure and temperature are largely independent of the conditions behind the shock. Because the dissociation results in a specific-heat increase and the enthalpy is essentially unchanged for dissociated or nondissociated gas, it means that the temperature decreases when dissociation is considered, which is expected because energy is absorbed in the dissociation process from the translational energy.

2.2.3 Reservoir Conditions and Thermal Choking in Constant-Area Ducts

The energy equation in equations of motion (2.24) can be conveniently written at a station where the velocity $u = 0$. These so-called reservoir conditions lead to the formulation of the stagnation-flow properties, of which the temperature is given for a perfect gas as

$$T_0 = T + \frac{u^2}{2c_p}. \tag{2.31}$$

Over a control volume that includes a portion of a duct that could represent, for example, a combustion chamber, in which friction is neglected and heat is added to the flow, the equations of motion, along with the equation of state, can be written in a differential form as

$$\frac{d\rho}{\rho} + \frac{du}{u} = 0,$$

$$\frac{dp}{p} + \gamma M^2 \frac{du}{u} = 0, \tag{2.32}$$

$$\frac{dT_0}{T} = \frac{dT}{T} + (\gamma - 1)M^2 \frac{du}{u}$$

or

$$\frac{dT_0}{T_0} = \frac{dT}{T} + \frac{(\gamma - 1)M\,dM}{\left(1 + \frac{\gamma - 1}{2}M^2\right)}.$$

In Eqs. (2.32) the Mach number dependency appears by use of the definitions in Eqs. (2.26) and includes the isentropic assumption expressed in the formulation of the speed of sound. The energy equation has been replaced with the stagnation-temperature definition, which changes because of the heat added to the system as $dT_0 = dq/(\rho u c_p)$.

The equation of state, written as

$$\frac{dp}{p} - \frac{d\rho}{\rho} - \frac{dT}{T} = 0, \tag{2.33}$$

completes the system that, after algebraic rearrangement, can be brought to a form that emphasizes the Mach number dependence on the stagnation-temperature change:

$$\frac{dM}{M} \frac{2(1 - M^2)}{(1 + \gamma M^2)\left(1 + \frac{\gamma - 1}{2}M^2\right)} = \frac{dT_0}{T_0}. \tag{2.34}$$

Equation (2.34), which is integrated in Figure 2.2, shows that heat addition to a frictionless flow in a duct causes the Mach number to approach unity for both subsonic and supersonic cases. For the scramjet engine, the implication is that thermal choking is reached after the delivery of a certain amount of heat to the flow. Further heat addition is possible only when accompanied by a reduction in mass flow and therefore changing the upstream conditions. Figure 2.2 shows the result of integrating Eq. (2.34) with the temperature normalized by its value corresponding to Mach = 1. Noticeable in the figure is the

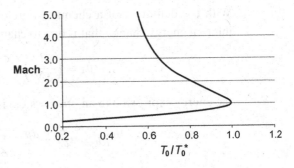

Figure 2.2. Thermal choking leads the Mach number toward unity, with a faster slope on the supersonic side.

more abrupt slope on the supersonic branch in comparison with the subsonic conditions. It is also clear in the figure that, for the supersonic branch, the Mach number drop is faster when starting from an initial higher Mach number for the same amount of heat deposition in the flow.

2.3 Chemical Reactions and Equilibrium

Chemically reactions accompany power-generating flows, and the system obeys the thermodynamic relations that apply equally to systems that are non-reacting or considered stationary. The condition of equilibrium, which is often invoked for simplicity, implies that processes taking place at the molecular level, such as chemical reactions, proceed with infinite speed and the gas adapts instantaneously to changes in its composition. The state variables used in the description of the flows through the conservation equations can be derived then from knowledge of the chemical composition. The following subsection includes the derivation of the equations that are used for chemically reacting flows, with the equilibrium assumption used to determine the gas composition.

2.3.1 Thermodynamic Relations and the Gibbs Function

In many combustion problems involving flow and chemical reactions, the assumptions of equilibrium hold and can therefore be described through thermodynamic relations. These relations are often applicable to cases that depart slightly from equilibrium; in those cases the derivation is based on simplified assumptions. When chemical reactions take place, in addition to the intensive properties, pressure and temperature, used in simplified flow analyses in the previous sections, the species-concentration change must be included. Assuming steady state and neglecting transport by diffusion, we find that the species concentration changes in response only to production that results from the chemical reactions. Conservation of species equation (2.11) thus changes to

$$\bar{u} \cdot \nabla Y^i = \frac{w^i}{\rho}, \quad i = 1, \ldots, N. \tag{2.35}$$

With the definition of a chemical potential of a given species reflecting the internal energy change that is due to changes in species molar numbers,

$$\mu_i \equiv \left(\frac{\partial U}{\partial N_i} \right)_{S, \rho, N_k}, \quad k \neq i, \tag{2.36}$$

the energy expression in an open system is

$$dh = T ds + \frac{dp}{\rho} + \sum_{i=1}^{N} \left(\frac{\mu_i}{M_i} \right) dY_i, \tag{2.37}$$

where M_i is the molecular weight of species i. These are thus a system of $N + 5$ differential equations involving $N + 5$ thermodynamic unknown properties, p, $\bar{u}$, T, and Y_i. For 1D flow both the number of equations and the number of unknowns reduce to $N + 3$. Density is related to pressure and temperature through the equation of state.

The Gibbs function, which is a state property defined as $G \equiv H - TS$, takes the form

$$dg = -S dT + \frac{dp}{\rho} + \sum_{i=1}^{N} \left(\frac{\mu_i}{M_i} \right) dY_i \tag{2.38}$$

when expressed per unit mole. This formulation of the Gibbs function is applicable to open, reacting systems. Because the Gibbs function is an extensive property, we can integrate Eq. (2.38) by assuming a system in which the extensive properties change while all species maintain the same intensive properties. Hence, because T, p, and μ_i remain constant, integration of Eq. (2.38) results in

$$g = \sum_{i=1}^{N} \frac{\mu_i}{M_i} Y_i. \tag{2.39}$$

2.3.2 Chemical Equilibrium

For an open system, Eq. (2.37) indicates that the enthalpy is expressed as the sum of heat transferred to the system added to the work done on its boundaries. The second law of thermodynamics defines the entropy as an extensive property that will be maximized at equilibrium conditions; hence, in a closed system that is not subject to internal constraints, the entropy change over an infinitesimal process will equal or exceed the ratio of heat addition to the equilibrium temperature:

$$ds \geq \frac{dq}{T}. \tag{2.40}$$

The equal sign holds for reversible processes and the inequality holds for all natural processes. Using the first law of thermodynamics, along with relation (2.40) and substitution into Eq. (2.37), we find that

$$\sum_{i=1}^{N} \frac{\mu_i}{M_i} dY_i \leq 0. \tag{2.41}$$

The inequality in this expression indicates that for a chemically reacting system the reactions move in the direction of decreasing chemical potential. Equilibrium will be reached, then, when the potential of the reactants equals that of the products (Glassman, 1996). Relation (2.41) makes evident the significance of the Gibbs function in defining chemical equilibrium because the existence of equilibrium implies that

$$dg|_{T,p} = 0 \tag{2.42}$$

when the temperature T and the pressure p are assumed constant. A single chemical reaction may be written in the form

$$\sum_{i=1}^{N} v_i' R_i \rightleftarrows \sum_{j=1}^{M} v_j'' P_j, \tag{2.43}$$

where v_i' are the stoichiometric coefficients of an arbitrary set of reactants R_i that are in equilibrium with the products P_j with stoichiometric coefficients v_j''. The stoichiometric coefficients balance the atomic species in the chemical-reaction equation. Because the Gibbs function is an extensive property, the total free energy of the mixture including the reactants and the products in reaction (2.43) assumed in equilibrium will be the sum of individual free energies:

$$G = \sum n_i G_i, \quad i = R_1, R_2, \ldots, R_i, P_1, P_2, \ldots, P_j, \tag{2.44}$$

where n_i represents the instantaneous number of moles of each compound in the mixture. Because the definition of the Gibbs function relates it to enthalpy and entropy, and because these two state variables can be defined with respect to a standard condition, which is most often taken at $p = 1$ atm, the Gibbs function can also be defined with respect to its value at a standard condition as

$$G_i(T, p) = G_i^0 + RT \ln(p_i/p_0), \tag{2.45}$$

where G_i^0 is the free energy at the standard state and p_i is the partial pressure of species i representing the contribution of each species to the total pressure $p_i = N_i p$, where N_i is the mole fraction of species i. Following an additional derivation (see, for example, Glassman, 1996), a relation between the partial pressure of the reactants and the products entering chemical reaction (2.43) is obtained as

$$-\Delta G^0 = RT \ln \left(\prod_j P_j^{v_j''} \Big/ \prod_i R_i^{v_i'} \right). \tag{2.46}$$

Here, the partial pressures of the species in the mixture are related to the standard Gibbs function. The ratio of the partial pressures in Eq. (2.46) is defined as the equilibrium constant,

$$K_p \equiv \prod_j P_j^{v_j''} \Big/ \prod_i R_i^{v_i'}, \tag{2.47}$$

which is determined from

$$K_p = \exp(-\Delta G^0 / RT). \tag{2.48}$$

The equilibrium constant K_p is thus a function of temperature. Based on the values of equilibrium reactions of formation of various species, the values of equilibrium constants of formation have been calculated for a large set of compounds and are tabulated, for example, in the NIST-JANNAF (1998) tables (NIST is the U.S. National Institute of Standards and Technology and JANNAF is the Joint Army–Navy–NASA–Air Force).

Equilibrium flow is a limiting case when the reaction rates are assumed to propagate much faster than the flow time scale. This is an assumption that can be applied to certain practical systems, for example, in certain regions of a solid-fuel rocket combustion chamber (Kuo et al., 1984) or in other applications in which the flows are moving slowly. The other limiting case is represented by the assumption that the fluid time scales are significantly shorter than the reaction rates; therefore the reactions do not propagate to any significant degree within the time scale of interest of the flow. This case is known as *frozen flow*, when the chemical reaction rates ω, which are discussed in the following subsection, become negligible; in other words, the chemical-reaction characteristic time becomes very long. Because the gas composition does not change, the implication is that the ratio of specific heats γ is considered constant and the gas can be treated as ideal.

2.3.3 The Law of Mass Action and Reaction-Rate Constants

If we consider again the forward-propagating reaction in reversible system (2.43), any change in the reactants' concentration will result in a comparable change in the products' concentration (Glassman, 1996; Williams, 1985). In other words, if n_i is the reactants' concentration, with $n_i = N_i N_A$, where N_i is the number of moles of species i and N_A is Avogadro's number, 6.022×10^{23} 1/mol, the time rate of change of species I concentration will result from chemical reaction (2.43) as

$$\frac{dn_i}{dt} \Big/ (v_i'' - v_i') = \frac{dn_j}{dt} \Big/ (v_j'' - v_j'). \tag{2.49}$$

This equality evidences the existence of a parameter that defines the reaction rate:

$$\omega \equiv \frac{dn_i}{dt} / (v_i'' - v_i'), \quad i = 1, \ldots, N. \tag{2.50}$$

The law of mass action relates the reaction rate ω to the concentration of the reactants:

$$\omega = k \prod_{i=1}^{N} (n_i)^{v_i'}, \tag{2.51}$$

where k is the specific reaction-rate constant, defined in most cases in an Arrhenius form, $k = AT^n \exp(-E/RT)$, where A, n, and E are determined from experiments. It should be noted that the derivation of the reaction rate in Eq. (2.50) corresponds to a single-step, forward-propagating reaction. In general, most reactions involve multiple elementary steps and the rate of reaction for a given species i results as the sum of the elementary reactions rates that characterize each individual step (Williams, 1985). With similar reasoning, for reactions that propagate in both directions, such as reaction (2.43), a backward-propagating reaction rate can be defined along with the forward-propagating reaction rate. In this case, the reaction rate will depend on both forward and backward specific reaction-rate constants:

$$\omega_{f \rightleftarrows b} = k_f \prod_{i=1}^{N} (n_i)^{v_i'} - k_b \prod_{j=1}^{N} (n_j)^{v_j'}, \tag{2.52}$$

where k_f and k_b represent the forward and the backward specific reaction-rate constants, respectively. The quantity $\omega_{f \rightleftarrows b}$ is the net reaction rate for reaction (2.43). At equilibrium, the net production of species is zero; therefore, from Eq. (2.52), the result is that a relation between the forward and backward specific reaction rates exists in the form

$$\frac{k_f}{k_b} = \frac{\prod_{j=1}^{N} n_j''}{\prod_{i=1}^{N} n_i'} \equiv K_n, \tag{2.53}$$

where K_n is the equilibrium constant written in terms of concentration and is related to the equilibrium constant determined in Subsection 2.3.2 through the relation between concentration and partial pressures, $p_i = n_i p$:

$$K_p = K_n \left(p / \sum n_i \right)^{\left(\sum v_j'' - \sum v_i' \right)}. \tag{2.54}$$

Equations (2.53) and (2.54) allow the calculation of one specific reaction-rate constant from the other by use of the equilibrium constant.

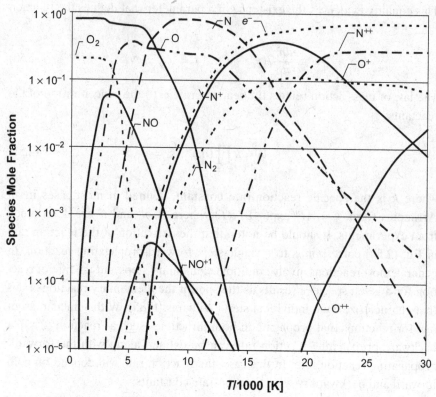

Figure 2.3. Air equilibrium composition for temperatures up to 30 000 K. Below 2000 K there is no significant dissociation at this pressure.

2.3.4 Air Equilibrium Composition

The temperature rise associated with high-speed flight causes dissociation of molecular oxygen and even nitrogen, resulting in a composition that includes atomic species along with ionized oxygen and nitrogen. Additional species appear at elevated temperatures as a result of the recombination of atomic species. Because significant nitrogen dissociation occurs at temperatures in excess of 5000 K whereas oxygen is almost entirely dissociated at that temperature, it is often assumed (Hansen 1959) that the dissociation reactions of these species occur independently. Thus simple models of air composition, such as those proposed by Hansen (1959), include only reactions that result in dissociation of molecular species, i.e., oxygen and nitrogen, and ionization of the atomic species:

$$
\begin{aligned}
O_2 &\to 2O, \\
N_2 &\to 2N, \\
O &\to O^+ + e^-, \\
N &\to N^+ + e^-,
\end{aligned}
\tag{2.55}
$$

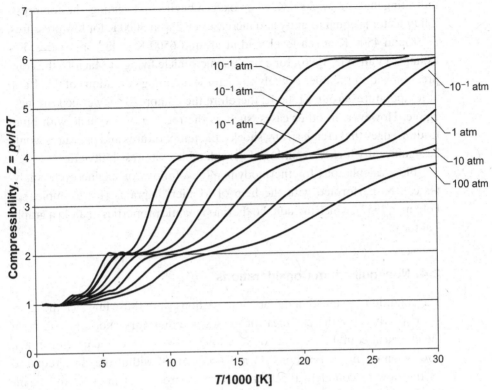

Figure 2.4. Compressibility of air as a function of temperature and pressure (after Thompson, 1991).

This model neglects additional species, among which NO is notable, with as much as 30% mole fraction around 3000 K at 1 atm (Hilsenrath et al., 1959). Figure 2.3 shows the gas composition as a function of temperature based on the model by Gupta et al. (1991), which assumes the presence of 11 species: O_2, N_2, O, N, NO, NO^+, O^+, O^{++}, N^+, N^{++}, e. The model by Hilsenrath et al. (1959) assumes the presence of 28 species including, in addition to the species in the Gupta et al. (1991) model, Ar, CO, CO_2, Ne, and ions of these species.

We can account for the departure from the air composition in the mixture at equilibrium by including the compressibility factor Z in the equation of state, continuing to consider the mixture as thermally perfect:

$$p = Z\rho RT, \tag{2.56}$$

where Z is the number of moles of mixture formed from a mole of the initially undissociated air. Thus the compressibility factor $Z = M_0/\overline{M}$, where M_0 is the air molecular weight and $\overline{M}$ is the mean molecular weight of the mixture. Figure 2.4, taken from Thompson et al. (1991), shows the effect of the compressibility factor on the equation of state for thermally perfect air, indicating that, below 2000 K, when oxygen begins dissociating, the compressibility factor is equal to unity and increases by 2% at 3000 K for low pressures, at around 4500 K at sea level, and at around 6500 K at high pressures. For example (Van Wie, 1996), for a hypersonic vehicle flying at Mach 10 the compression within the inlet is likely to increase, creating conditions of 0.5–1 atm at temperatures around 1500 K; therefore the compressibility effects are negligible. However, in other cases, such as the reentry of a vehicle with blunt leading edges that create strong shocks, the temperatures and pressures would be significantly higher, resulting in considerable compressibility effects.

Flow calculations for thermodynamic evolutions or engine cycle analyses can be performed with the help of Mollier diagrams (for example, see Feldman, 1957), which include the thermodynamic properties of air in a graphical form.

2.4. Nonequilibrium Considerations

The conditions assumed in the previous sections were those in which the characteristic time for the chemical processes is either very short, $\tau \rightarrow 0$, as in the chemical-equilibrium case, or very long, $\tau \rightarrow \infty$, as in the case of frozen flow when chemical reactions do not occur at all within the time required by the flow to complete the thermodynamic process of interest, for example, expansion through a nozzle. The condition for equilibrium requires an internal energy adjustment over the molecular-level states, and it is realized through molecular collisions. When the time required for certain molecular processes becomes comparable with the fluid dynamics time scales, the condition of equilibrium cannot be satisfied. Certain processes that lead to nonequilibrium, such as transport phenomena, proceed at high speed and therefore appear only in flows with large gradients, such as a thin layer separating a low-speed region from a high-speed region. An example of such a situation is the boundary between a low-speed, high-temperature region in which combustion takes place and the adjacent high-speed, nonreacting flow that may coexist in the combustion chamber in a scramjet. It has been estimated (Segal et al., 1995) that in a case of combustion of a hydrogen transverse jet in a Mach 2 flow the characteristic time for combustion is 12–16 μs whereas the average velocity across the shear layer formed downstream of the jet plume is of the order of 50 μs. In particular, because of the large velocity and thermal gradients across this shear layer, these time scales indicate that nonequilibrium effects must be taken into account. The chemical reactions must therefore be treated as finite. In certain cases, as previously mentioned, the transport processes must also be treated as proceeding at a finite rate. A discussion of this treatment, from a molecular-level point of view, is given by Vinceti and Kruger (1986). Other nonequilibrium effects are due to processes that are significantly longer, such as vibrational relaxation, or to chemical reactions that proceed at a finite rate as described in Subsection 2.3.3. Because these processes are of relatively long duration, their nonequilibrium effects are significant even at lower temperatures and flow gradients.

We can write a complete set of equations for nonequilibrium flow by deriving a population-change time rate for atoms and molecules at a specific energy level i, e.g., the difference between the sum of all rates of collisional and the much slower radiative transitions that populate a given state and the sum of the rates that depopulate the same state (Cheng and Emanuel, 1995). The transitions for each collisional process are determined for molecular, atomic, and electronic interactions by use of the relations derived from quantum mechanics.

REFERENCES

Batchelor, G. K. (1994). *An Introduction to Fluid Dynamics*, Cambridge University Press.

Cheng, H. K. and Emanuel, G. (1995). "Perspective on hypersonic nonequilibrium flow," *AIAA J.* **33**, 385–400.

Cullen, H. B. (1985). *Thermodynamics and an Introduction to Thermostatistics*, 2nd ed., Wiley.

Feldman, S. (1957). "Hypersonic gas dynamic charts for equilibrium air," AVCO Research Laboratories.

Glassman, I. (1996). *Combustion*, 3rd ed., Academic Press.

Gupta, R. N., Lee, K.-P., Thompson, R. A., and Yos, J. M. (1991). "Calculations and curve fits of thermodynamic and transport properties for equilibrium air to 30000 K," NASA RP-1260.

Hansen, C. F. (1959). "Approximations for the thermodynamic and transport properties of high-temperature air," NASA TR R-50.

Hilsenrath, J., Klein, M., and Woolley, H. W. (1959). "Tables of thermodynamic properties of air including dissociation and ionization from 1500 K to 15000 K," Arnold Engineering Development Center, AEDC-TR-59-20.

Incropera, F. P. and DeWitt, D. P. (2002). *Fundamentals of Heat and Mass Transfer*, 5th ed., Wiley.

Kuo, K. K., Gore, J. P., and Summerfield, M. (1984). "Transient burning of solid propellants," in *Fundamentals of Solid-Propellant Combustion* (K. K. Kuo and M. Summerfield, eds.), Progress in Astronautics and Aeronautics, AIAA, Vol. 90, pp. 599–659.

Liepmann, H. W. and Roshko, A. (1957). *Elements of Gasdynamics*, 2nd ed., Wiley.

NIST-JANNAF Thermochemical Tables. (1998). *J. Phys. Chem. Ref. Data*, 4th ed.

Segal, C., Krauss, R. H., Whitehurst, R. B., and McDaniel, J. C. (1995). "Mixing and chemical kinetics interactions in a Mach 2 reacting flow," *J. Propul. Power* **11**, 308–314.

Thompson, R. A., Lee, K.-P., and Gupta, R. N. (1991). "Computer codes for the evaluation of thermodynamic and transport properties for equilibrium air to 30000 K," NASA TM-104107.

Van Wie, D. M. (1996). "Internal flowfield characteristics of a scramjet inlet at Mach 10," *J. Propul. Power* **12**, 158–164.

Vincenti, W. G. and Kruger, C. H. (1986). *Introduction to Physical Gas Dynamics*, Krieger.

Williams, F. A. (1985). *Combustion Theory*, Addison-Wesley.

3 High-Temperature Gas Dynamics and Hypersonic Effects

3.1 Introduction

As the air energy increases, molecular vibrational excitation, followed by dissociation and chemical reactions, accompanies flows at high speed; it occurs on both the external vehicle surface and in the scramjet flow path. The viscous layer generated on the vehicle forebody, following deceleration and interactions with shock waves, may exhibit temperatures in excess of thousands of degrees. At these temperatures, chemical effects are just as important as in combustion chambers, where chemical reactions are intentionally induced and controlled; the gas composition departs substantially from the simplifications made through the assumption of thermal or calorically perfect gas and has a substantial impact on the flow structure, energy distribution, and, finally, thrust generation. The gas properties must therefore be determined from a microscale perspective that takes into account the molecular motion, the distribution, and transfer of energy between the molecules present in the flow.

This chapter reviews the molecular motion and the determination of real-gas and transport properties from the description of the flow molecular structure. Further, specific issues that distinguish the high-speed flows in the regime that is generically referred to as "hypersonic" are reviewed. Without any abrupt demarcations in the flow properties, the hypersonic regime is considered in general to exist when the flight Mach number exceeds a value of ~ 5. For a more accurate definition, the hypersonic regime could be considered to manifest when the entropy layer close to a surface immersed in the fluid becomes significant (Anderson, 1989). Although the flows through a scramjet approach this limit toward only the upper limit of its operability (see the cycle analysis discussion in Chap. 4), when the flow is dominated by heat release and large velocity and thermal gradients, the external hypersonic aerodynamics are of substantial significance, given the strong coupling between the vehicle and the propulsion system for this flight regime. The hypersonic shock waves are

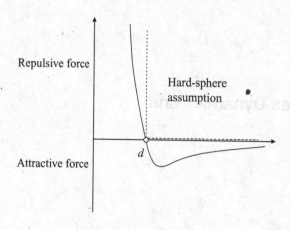

Figure 3.1. Intermolecular force dependence on the distance between two molecules.

very close to the surface, and therefore the entire forebody of a hypersonic vehicle becomes part of the engine compression system. The shock detachment point at the vehicle's leading edge and the fluid and thermodynamic features that ensue for this thin layer have a significant influence on the processes that take place in the scramjet engine. For this reason a brief discussion of the main features of hypersonic flows is included here.

3.2 Real-Gas Equation of State

The thermally perfect gas assumes that the intermolecular forces and the molecule size are negligible. A schematic of the intermolecular force is shown in Fig. 3.1; it results from the electromagnetic field that is almost uniform around a molecule that is assumed to be spherically symmetric. Close up, the field is repulsive and increases rapidly as the electronic clouds come into contact with each other. As the molecules move apart from each other, an attractive force appears that becomes weaker. There is a distance therefore at which the intermolecular force is zero, and the molecules will maintain this equilibrium if the kinetic energy does not cause them to move from this position. The assumption of perfect gas translates into the assumption of rigid spherical molecules for which an infinite intermolecular force exists below a distance d that represents the diameter of the spherical molecule and a zero intermolecular for distances larger than d.

At low temperatures and high pressures, the molecules are close to each other and the manifestation of real-gas behavior becomes important. It is generally accepted that, for pressures around 1000 atm or temperatures below 30 K (Anderson, 1989), real-gas formulations must be used instead of simplified assumptions. Intermolecular forces depend on both pressure and temperature, and deviations from the perfect-gas assumption scale with p/T^3. In general, however, for most practical applications the perfect-gas assumption is acceptable.

The van der Waals equation of state is widely used:

$$p = \frac{RT}{v - b} - \frac{a}{v^2}. \tag{3.1}$$

Here, a and b are constants that are specific to the particular gas, v is the specific volume, and R is the universal gas constant, 8.314 Nm/(mole K). The first constant, a, takes into account the presence of intermolecular forces that modify the pressure as a molecule approaching the wall of a container is attracted back by the other molecules in the volume (Cullen, 1985). Constant b reflects a reduction of the volume occupied by the gas that is due to the nonzero dimension of the molecules. The van der Waals equation of state has a good qualitative ability to represent liquid–gas phase transition (Emanuel, 1987), although other equations of state for real gases predict with greater accuracy the behavior of real gases. Other equations of state have been suggested to predict real gas behavior, among them the Redlich–Kwong equation,

$$p = \frac{RT}{v - b} - \frac{a}{T^{1/2}(v^2 + vb)}, \tag{3.2}$$

and the virial form of the equation of state,

$$\frac{pv}{RT} = 1 + \frac{B(T)}{v} + \frac{C(T)}{v^2} + \cdots +, \tag{3.3}$$

where $B(T), C(T), \ldots$, are called the second, third,..., virial coefficients. For high pressures and temperatures approaching the thermodynamic critical conditions, the Peng–Robinson–Stryjek–Vera (PRSV) equation of state gives accurate solutions:

$$p = RT(V - b) - a\alpha[V(V + b) + b(V - b)], \tag{3.4}$$

where a, b, and α are parameters specific to each fluid.

3.3 Elements of Kinetic Theory

3.3.1 Pressure, Energy, and the Equation of State

Assuming a simplified model of the molecules as "hard spheres," in other words, nondeformable, spherical solids of specified diameter and mass, a model of the speed associated with the random motion leads to the description of a number of gas properties including pressure, temperature, and internal energy. Other properties can be determined from knowledge of the internal structure of the molecule.

Considering a gas in a state of equilibrium contained in a motionless box, the molecules' motion inside the box is random. The pressure exerted by the gas on the box walls is a direct effect of this random motion. The collisions

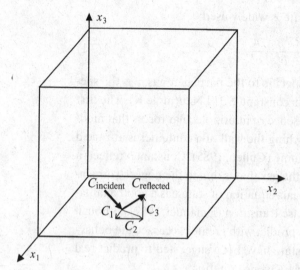

Figure 3.2. The reflection of a molecule in a cubical box is specular, i.e., the reflection angle equals the incidence angle and the velocity magnitude is maintained.

between molecules can be neglected because of the assumption of equilibrium, which implies that, for each molecule changing direction as a result of an intermolecular collision, another molecule would move with the same speed in the exact opposite direction. Similarly, when a molecule hits a wall, the reflection must be, under the same equilibrium assumption, specular, i.e., the angle of reflection equals the angle of incidence.

Assuming the chosen box is of cubical shape, Fig. 3.2 shows the motion of the molecule that has reached the wall at point P with a velocity C. It reflects specularly, meaning that the magnitude of the velocity components $|C_1|$, $|C_2|$, $|C_3|$ are maintained after the reflection, whereas the directions of the velocity components perpendicular on the wall become opposite and the directions of the components parallel to the wall remain unchanged.

The force exerted by the molecule on the box wall depends on the number of collisions with the wall. The number of collisions of a molecule per unit time with wall (x_2, x_3) is $|C_1|/2l$, where l is the box length and the ratio indicates that it transverses back and forth between two consecutive collisions with the same wall. The change in momentum experienced by the molecule with each collision is $2m|C_1|$, where m is the molecule's mass. The force exerted on the wall by the molecule is thus equal to the change in momentum experienced by the molecule during the collision:

$$2m|C_1| \; |C_1|/2l = mC_1^2/l. \tag{3.5}$$

Because pressure results from force applied to a surface, the pressure exerted by one molecule in one direction is mC_1^2/V, where V is the volume of the box. A mixture of molecules with masses m_i may be present in the box, and they exert pressure on all three faces of the box. Therefore the pressure in the box will be the average of the contributions of all molecules colliding with the box walls in all directions; thus,

$$p = \frac{1}{3V} \sum_i m_i \left(C_{i,1}^2 + C_{i,2}^2 + C_{i,3}^2 \right) = \frac{1}{3V} \sum_i m_i C_i^2. \tag{3.6}$$

The kinetic energy associated with the translational motion of the molecules is

$$E_{\mathrm{tr}} = \frac{1}{2} \sum_i m_i C_i^2, \tag{3.7}$$

which, when combined with Eq. (3.6), results in

$$pV = \frac{2}{3} E_{\mathrm{tr}}. \tag{3.8}$$

Equation (3.8) is the kinetic theory equation of state. When compared with the equation of state derived from thermodynamic considerations, $pV = NRT$, where N is the number of moles in the system, the kinetic energy for translational motion becomes directly related to the temperature of the system:

$$E_{\mathrm{tr}} = \frac{3}{2} RT. \tag{3.9}$$

For a single molecule the kinetic energy is

$$\hat{e}_{\mathrm{tr}} = \frac{E_{\mathrm{tr}}}{N} = \frac{3}{2} \frac{NRT}{N} = \frac{3}{2} kT, \tag{3.10}$$

where $k = R/\hat{N}$ is the Boltzmann constant, $1.38 \times 10^{-23} J/K$, and $\hat{N}$ is the number of molecules per mole or Avogadro's number, 6.022×10^{23} mol^{-1}. In terms of R, the gas constant, the kinetic energy of translation per unit mass is

$$e_{\mathrm{tr}} = \frac{3}{2} RT. \tag{3.11}$$

Considering a monoatomic molecule at temperatures sufficiently low to preclude ionization, without disturbing the internal structure, the kinetic energy of translation is the only form of internal energy that the molecules can possess. The specific heat at constant volume is thus

$$c_v = \left(\frac{\partial e}{\partial T} \right)_v = \frac{3}{2} R. \tag{3.12}$$

Because the specific heat at constant pressure is $c_p - c_v = R$ and the ratio of the specific heats is $\gamma \equiv c_p/c_v$, for the monoatomic gas,

$$c_p = \frac{5}{2} R, \quad \gamma = \frac{5}{3}. \tag{3.13}$$

The model assumed in this discussion is both thermally and calorically perfect and therefore describes a perfect gas.

In the derivation of the molecular energy, in Eq. (3.10), the only variable is temperature; therefore there is no distinction between molecules with different molecular weights. In other words, all molecules in this model have the same molecular kinetic energy irrespective of the molecular weight. This indicates that, on average, the heavier molecules move slower than the lighter molecules in a mixture kept at the same temperature.

A mean-square velocity can be defined by substitution of the pressure defined in Eq. (3.6) from kinetic consideration into the equation of state:

$$\bar{C}^2 \equiv \frac{\sum_i m_i C_i^2}{\sum_i m_i}, \tag{3.14}$$

which, when combined with the equation of state, results in an expression for the root-mean-square speed:

$$\sqrt{\bar{C}^2} = \sqrt{3\frac{p}{\rho}} = \sqrt{3RT}. \tag{3.15}$$

Hence the molecular speed can be found from the macroscopic values for pressure and density. For example, for helium at standard conditions, the density is $\rho = 0.162$ kg/m^3. At a pressure of 1 atm, the root-mean-square molecular speed is found as $\sqrt{\bar{C}^2} = 1360$ m/s, which is of the same order as the speed of sound, $a = 1020$ m/s, which is expected because sound propagation is related to molecular motion.

3.3.2 Mean Free Path

The previous discussion did not include the collision between molecules. A concept of significant importance in the analysis of gases at high temperature is the *mean free path*, which is defined as the average distance a molecule travels between two successive collisions. The mean free path helps identify whether the gas can be treated as a continuum or if the molecules are sufficiently spread apart; in other words, the gas is rarefied to the point at which the continuum approximation cannot be invoked and the gas has to be treated as a collection of individual molecules.

Considering a gas formed from spherical molecules of the same diameter d, a molecule A will collide with another molecule B when it approaches the center of B at a distance equal to its diameter d. The situation is shown in Fig. 3.3. In its motion, molecule A will interact with another molecule whose center approaches the center of molecule A at a distance d. Therefore molecule A is said to generate in its motion a *sphere of influence*. As it travels at the average speed $\bar{C}$, the sphere of influence sweeps a cylindrical volume equal to $\pi d^2 \bar{C}$. If n is the number of molecules per unit volume, i.e., the number

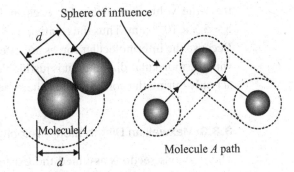

Figure 3.3. Path of a molecule A and its sphere of influence.

density, there will be $n\pi d^2\bar{C}$ collisions per second. This quantity is referred to as the *collision frequency* for molecule A:

$$Z_A = n\pi d^2\bar{C}. \tag{3.16}$$

The average distance traveled by molecule A between two consecutive collisions is the *mean free path*:

$$\lambda = \frac{\bar{C}}{Z_A} = \frac{1}{n\pi d^2} = \frac{m}{\rho\pi d^2}, \tag{3.17}$$

where the mass of the molecules m and the density ρ are also included. This derivation assumes that the velocity of a single molecule relative to the rest of the molecules present in the volume is considered motionless. When the relative speed of the molecules is taken into account, the relative velocity is $\sqrt{2}C$ and the mean-free-path equation becomes

$$\lambda = \frac{1}{\sqrt{2}n\pi d^2}. \tag{3.18}$$

Because the mean free path changes with the inverse of the number density, it means that it is proportional to the temperature and inversely proportional to the pressure, i.e., $\lambda \propto T/p$. As a result, in high-temperature flows at low pressures, the separation between consecutive collisions can be relatively large, and therefore particularly in high-speed flows, the time required for completing chemical reactions may become comparable with the residence time and the condition of equilibrium may not be met. The mean free path is related to a characteristic dimension in the system appropriately selected, for example, the height of a two-dimensional combustion chamber H, through the *Knudsen number*, $\mathrm{Kn} = \lambda/H$. If $\mathrm{Kn} \geq 1$, the flow is called free-molecule flow and the continuum treatment of the gas needs to be modified. On Earth, at sea level, a cubic millimeter contains approximately 2.5×10^{16} molecules. A typical molecule diameter is 3×10^{-4} μm and the mean free path is of the order of 10^{-2} μm, which is significantly smaller than most characteristic lengths in problems of practical interest. The spacing between molecules δ is given from the volume available for each molecule, i.e., $\delta = 1/n^3$, and therefore $\delta \approx 3.3 \times 10^{-6}$ mm. Thus the relative molecular size scales as $\lambda{:}\delta{:}d \approx 100{:}10{:}1$. Because the intermolecular forces are effective only for a distance comparable with the molecular diameter, it is justifiable to assume that the molecules interact only during the collision and otherwise neglect the intermolecular forces.

3.3.3 Maxwellian Distribution – Velocity Distribution Function

The previous sections assumed the existence of a single velocity $\bar{C}$ of a single molecule between two consecutive collisions. In reality, not all molecules travel with the same velocity, and the velocity of each given molecule changes with time. The existence of this broad range of velocities is accounted for through a *velocity distribution function* $f(C_i)$, which is a probabilistic treatment of the velocity representing the total number of particles with velocities within a velocity space, $dC_x dC_y dC_z$. Thus the velocity distribution function satisfies the condition that

$$\int_{-\infty}^{\infty} f(C_i)dV_c = 1. \tag{3.19}$$

A derivation of the velocity distribution function can be found in Vincenti and Kruger (1986). The result for equilibrium velocity, first derived by Maxwell in the 19th century, is

$$f(C_i) = f(C_x, C_y, C_z) = \left(\frac{m}{2\pi kT}\right)^{3/2} \exp\left[-\frac{m}{2kT}\left(C_x^2 + C_y^2 + C_z^2\right)\right]. \tag{3.20}$$

This equation gives the number of particles in a velocity space, known as the Maxwellian distribution. In a system at equilibrium, the function is symmetric with respect of its velocity components in the sense that, for each molecule leaving the velocity range of interest in any given direction, another will enter the same range. Thus, at equilibrium, the direction of the velocity vectors is not relevant; the magnitude of the velocity vectors is of interest. Therefore, by integrating the velocity distribution function within an interval of velocities $C + dC$, as depicted in the velocity space shown in Fig. 3.4, we obtain the speed distribution function as

$$\chi(C) = 4\pi \left(\frac{m}{2\pi kT}\right)^{3/2} C^2 \exp\left(-\frac{mC^2}{2kT}\right). \tag{3.21}$$

When plotted as a function of velocity, as shown in Fig. 3.5, the speed distribution function identifies certain speeds of particular interest:

1. The most probable speed determined by the maximum of the curve. From differentiation this speed is

$$C_{mp} = \sqrt{2kT/m}. \tag{3.22}$$

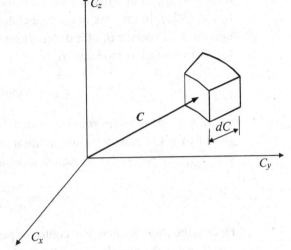

Figure 3.4. Volume element in the velocity space.

2. The average speed obtained from integration over the entire speed space, $\bar{C} = \int_0^\infty C \chi(C) \, dC$, resulting in

$$\bar{C} = \sqrt{\frac{8kT}{\pi m}}. \tag{3.23}$$

3. The root-mean-square speed obtained from $\bar{C}^2 = \int_0^\infty C^2 \chi(C) \, dC$ as

$$\sqrt{\bar{C}^2} = \sqrt{3kT/m}. \tag{3.24}$$

This is the same as the earlier result obtained in Eq. (3.15) from the discussion of pressure, energy, and the equation of state. Figure 3.5 indicates the relative values of the speeds just identified. By comparison, the speed of sound $a = \sqrt{\gamma kT/m}$ for a monoatomic gas has a magnitude of $a \simeq 0.9 \, C_{mp}$. It is interesting to note that all these speeds increase with temperature and decrease for heavier gases.

In this analysis, a single molecule's motion has been tracked. In reality, other molecules are present in the container and all will travel at different speeds. Furthermore, in mixtures, molecules of different sizes move with

Figure 3.5. The speed distribution function and the relative magnitude of the most probable mean and root-mean-square speeds.

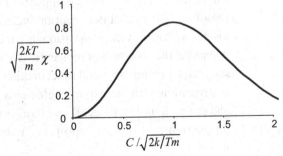

their own speed distribution function that, at equilibrium, can be described by Eq. (3.21). In the case of a single collision of species A molecules with the molecules of species B, of a different size, the collisional frequency given by Eq. (3.16) has to be modified to

$$Z_{AB} = n_B \pi d^2 \bar{C}_{AB},$$ \hfill (3.25)

where $\bar{C}_{AB}$ is the mean relative velocity between molecules A and B and is given by Eq. (3.23) with the observation that the molecular mass m is replaced with m^*_{AB}, the reduced mass, which is defined as

$$m^*_{AB} \equiv \frac{m_A m_B}{m_A + m_B}.$$ \hfill (3.26)

Finally, the total frequency of collisions per unit volume between molecules A and B, called the *bimolecular collisional rate*, is given by

$$Z_{AB} = n_A n_B d^2_{AB} \sqrt{\frac{8\pi kT}{m^*_{AB}}},$$ \hfill (3.27)

where d_{AB} is the radius of influence corresponding to a set of molecules A and B. It should be noted that the collisional rate between molecules of a single species cannot be obtained directly from Eq. (3.27) because the case of collision between two molecules of the same species but with different velocities C_1 and C_2 is indistinguishable from the case of the same molecules having velocities C_2 and C_1 and therefore the same collision would be counted twice; hence the $1/\sqrt{2}$ factor in Eq. (3.18).

3.3.4 Transport Coefficients

Gradients present in a fluid result in an exchange of mass, momentum, and energy from one area of the gas to another to reach equilibrium. These exchanges take place through molecular transport and are sensed at a macroscopic level through diffusion, viscosity, and heat conduction. The coefficient of heat transfer k appears in the Fourier law for conduction, given in Eq. (2.15), and the viscosity μ appears in the relation between the shear-stress tensor and the rate of deformation included in Eq. (2.16). For Newtonian fluids, the viscosity is the proportionality factor. The diffusion coefficient D_{AA} is defined as the transport of molecules of a gas A in a volume of molecules of the same kind per unit area and time; in other words, a coefficient of self-diffusion. A derivation of the transport coefficients is beyond the scope of this summary and can be found in Vincenti and Kruger (1986), which is based on accounting for the transport of a given property across a chosen boundary in the fluid. At moderate temperatures, the diffusion coefficients are found to depend on the molecular quantities as follows:

$$\mu = \beta_\mu \rho \bar{C} \lambda,$$

$$k = \beta_k \rho \bar{C} \lambda c_v, \tag{3.28}$$

$$D_{AA} = \beta_D \bar{C} \lambda,$$

where the coefficients β_μ, β_k, and β_D are numerical constants and λ is the mean free path defined in Eq. (3.16). It should be noted from Eqs. (3.28) that the molecular viscosity μ differs from the product ρD_{AA} through a constant. This constant is known as the *Schmidt number* and is used in the description and modeling of mixing processes. An evaluation of the transport properties for high-temperatures gases and mixtures is given by Mason (1969), who follows the derivations of Chapman and Cowling (1952) and Hirshfelder et al. (1954).

For multicomponent systems, such as chemically reacting flows, the mixture's transport values are found from individual species transport coefficients. The viscosity and thermal conductivity are then determined by use of Wilke's rule as

$$\mu = \sum_i \frac{X_i \mu_i}{\sum_j X_j \phi_j}, \tag{3.29}$$

with

$$\phi_{ij} = \frac{1}{\sqrt{8}} \left(1 + \frac{M_i}{M_j}\right)^{-1/2} \left[1 + \left(\frac{\mu_i}{\mu j}\right)^{1/2} \left(\frac{M_j}{M_i}\right)^{1/4}\right]^2, \tag{3.30}$$

where μ is the mixture viscosity, and μ_i, X_i, and M_i are components' i viscosities, mole fractions, and molecular weights, respectively. Wilke's rule applies in a similar expression to mixtures of thermal conductivities.

The problem of molecular diffusion at the interface of nonsimilar gases is of particular interest for air–fuel mixing in propulsion systems. If two parallel gaseous streams, of equal temperature and pressure but with different composition, travel with the same axial velocity, molecules from one gas will diffuse into the other with binary diffusion coefficients D_{12} and D_{21}. As diffusion progresses in an axial direction, the mixing layer that separates the single component streams becomes increasingly large. The flux of molecules from stream 1 that diffuses into stream 2 depends on the binary diffusion coefficient and the number density gradient as

$$\frac{\partial n_1}{\partial t} = -D_{12} \frac{\partial n_1}{\partial z}, \quad \frac{\partial n_2}{\partial t} = -D_{21} \frac{\partial n_2}{\partial z}, \tag{3.31}$$

where z is the direction perpendicular to the gases interface. Equations (3.31) are also known as *Fick's law for diffusion* (Eckert and Drake, 1959) and can be written in terms of the mass flux as

$$\dot{m}_1 = -D_{12}\frac{\partial \rho_1}{\partial z}. \tag{3.32}$$

Because the mixture's total number density is preserved, which means that $n_1 + n_2 = $ const, it follows that the number density gradients of the two gases are equal and of opposite sign, implying that $D_{12} = D_{21}$. The diffusion coefficients are largely inversely dependent on pressure and less sensitive to temperature and the gases' number densities (Chapman and Cowling, 1952).

3.4 Elements of Statistical Thermodynamics

The assumption of a calorically perfect gas, which predicts constant specific heat c_p, can lead to large errors as the temperature increases. For example, for a vehicle flying at 4000 m/s, the temperature based on the calorically perfect-gas assumption is $T_{CPG} = 8600$ K, whereas the actual temperature is half that value; during the Apollo lunar mission return, at a velocity of 11 000 m/s, the calculated $T_{CPG} = 58\,300$ K, whereas the actual temperature was of the order of 12 000 K. This indicates a substantial departure from the Hugoniot jump conditions for constant specific heat given by Eqs. (2.28), resulting in an almost linear dependence of temperature on the flight Mach number, $T \sim M$.

The major reasons for this departure from the calorically perfect-gas assumption are as follows:

- As the temperature increases, the vibrational motions in diatomic and polyatomic molecules increase, absorbing energy that would otherwise be distributed over the translation energy.
- A further increase in temperature leads to dissociation and even ionization, as shown in Fig. 2.4. The gas begins to change the chemical composition, and the specific heat becomes a function of both pressure and temperature, $c_p = c_p(p, T)$.

The diagram shown in Fig. 3.6 indicates the onset of dissociation for air at elevated temperatures.

Hence, for high temperatures, the thermodynamic properties, pressure and temperature, are more appropriately obtained from the equilibrium thermodynamic properties by means provided by *statistical thermodynamic theory*, which enables the calculation of macroscopic properties of a system from the microscopic particle structure that it contains. The relation between the microscopic structure at thermodynamic equilibrium and the macroscopic

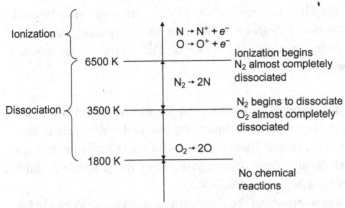

Figure 3.6. Dissociation and ionization of air begin approximately at 1800 and 6500 K, respectively.

properties is provided by the Boltzmann equation,

$$S = k \ln \Omega, \tag{3.33}$$

which relates the macroscopic value of entropy to a function Ω that describes the distribution of the particles that compose the system; in other words, the number of ways in which the microscopic particles can be arranged according to their properties and the macroscopic conditions.

3.4.1 Microscopic Description of Gases

3.4.1.1 Modes of Energy

In the following discussion, the system is assumed to be formed of a number of N identical molecules that interact with each other through collisions or other mechanisms, leading to an energy exchange. As a result, the particles will have different energetic states at microscopic levels. To analyze the forms of energy that a molecule may have, the following analysis treats a diatomic molecule simplified to a "dumbbell" model with two spheres connected by a 1D rod that represents the intermolecular force, as shown in Fig. 3.7.

Figure 3.7. Dumbbell representation of diatomic molecules in a 3D reference frame.

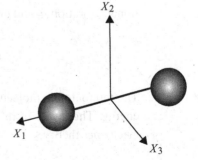

In this model the molecule may distribute its energy over translational, rotational, and vibrational degrees of freedom and may also accumulate energy in the electrons' motion in their orbits. These forms of energy are described as follows:

1. The *translational energy* ε'_{trans} results from the kinetic energy associated with the velocity vectors in three directions. The molecule is said to have three "degrees of freedom." These components of the kinetic energy manifest themselves through thermal energy; therefore the molecule is said to have three "thermal degrees of freedom."

2. The molecule can rotate about the three orthogonal axes, as shown in Fig. 3.7; therefore it possesses *rotational energy* ε'_{rot}. However, the moment of inertia associated with rotation around the axis that connects its atoms x_1 is negligible in comparison with the moments of inertia about the other two axes; therefore the molecule is said to possess only two "rotational degrees of freedom." The same is true for any linear polyatomic molecule, for example, the CO_2 molecule. Most of the polyatomic molecules do not have a linear structure and therefore exhibit all three rotational modes of energy.

3. The atoms vibrate with respect to each other about an equilibrium position. The diatomic model can be envisioned as having a spring connecting the two atoms. The potential energy accumulated in the spring during the vibration of the two atoms and the kinetic energy associated with the vibrational motion of the atoms contribute three "vibrational degrees of freedom" to the diatomic molecule's *vibrational energy* ε'_{vib}. Polyatomic molecules have a more complex vibrational structure that results in a large number of degrees of freedom.

4. The electrons moving along their orbits around the atomic nuclei contribute with two sources to the *electronic energy* ε'_{el}: the kinetic energy component and the potential energy in orbit resulting from the electrons' location in the electromagnetic field of the nucleus.

The total energy of the molecule is thus the sum of the translational, rotational, vibrational, and electronic energies:

$$\varepsilon' = \varepsilon'_{trans} + \varepsilon'_{rot} + \varepsilon'_{vib} + \varepsilon'_{el}. \qquad (3.34)$$

The monoatomic molecules do not have rotational and vibrational modes of energy. The levels of these energies have discrete values given by quantum mechanics theory.

3.4.1.2 Quantum Energy Levels and Degeneracies

Quantum mechanics results, starting from Heisenberg's uncertainty relation, conclude that a probability function can be formulated to indicate the presence of a particle in the vicinity of a precise position (Herzberg, 1989). This probability function, the *wave function* ψ, is defined such that the product $\psi^2 dV$ gives the probability that a given particle lies in a given physical element of space, dV. This description makes evident the wave nature of the small-particle motion and is derived as the solution to the *Schrödinger wave equation*:

$$\frac{h^2}{8\pi^2 m}\left(\frac{\partial^2 \psi}{\partial x_1^2} + \frac{\partial^2 \psi}{\partial x_2^2} + \frac{\partial^2 \psi}{\partial x_3^2}\right) + (\varepsilon - \varepsilon_p)\psi = 0, \tag{3.35}$$

where h is Planck's constant, 6.6256×10^{-27} erg s, m is the particle mass, ε is the total energy of the particle, and ε_p is its potential energy. The solutions of the Schrödinger wave equation emphasize the discrete nature of the energy levels that a particle can have; in other words, the energy is quantized at the microscopic level, whereas it appears continuous in the macroscopic classical analysis of motion.

The schematic of the energy levels shown in Fig. 3.8 emphasizes several features of the energy-mode levels. The height of each energy level represents its value in relation to the other energy modes. The lowest level for each energy mode, i.e., translational, rotational, vibrational, and electronic, is defined as the *ground* level. These levels are indicated by the subscript "0" with the subsequent higher energy levels numbered in ascending order. The ground levels correspond to the energy the molecules would have at absolute-zero temperature.

The translational energy modes are closely spaced, to the extent that the translational energy appears almost as continuous and, for simplicity, these modes are not included in the schematic in Fig. 3.8. The rotational energy spacing between adjacent levels is much larger, and furthermore the spacing increases as the energy increases. The spacing between adjacent vibrational energy levels is larger still than the rotational energy levels; however, unlike the rotational energy levels, the vibrational energy levels become more closely spaced as the energy level increases. Finally, the electronic energy levels are farther apart in comparison with the vibrational energy levels, and, like the latter, the difference between adjacent levels decreases as the energy increases.

In addition to the quantized translational, rotational, vibrational, and electronic energy levels that together render a certain quantized total energy level for the molecule, a distinction is made in quantum mechanics with regard to the orientation of the molecule. Thus the rotational momentum can have

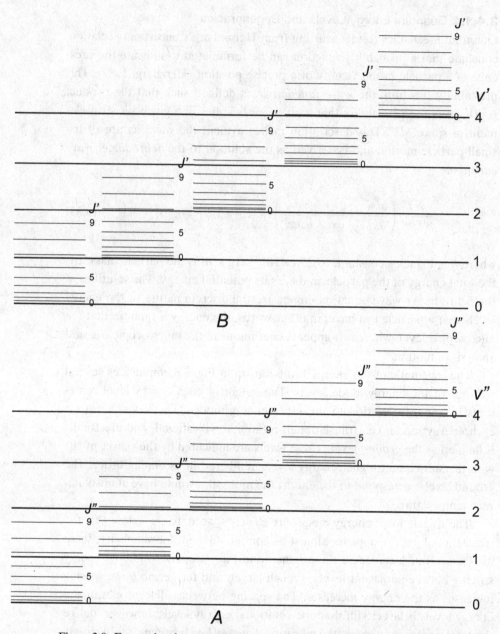

Figure 3.8. Energy-level spacing diagram. *A* and *B* are electronic levels with vibrational levels v' and v'' and rotational levels J' and J''. The translational levels are not shown due to the fine spacing that renders them almost as a continuum spectrum. The energy levels marked "0" represent the ground levels (after Herzberg, 1989).

only certain quantified directions, all with the same rotational energy level. Therefore there can be energy levels having several states that have the identical values of the energy ε_i, called, in this case, a *degenerate* level. The degeneracies are usually denoted by g_i and appear when the number and distribution of molecules are counted over the available energy levels.

3.4.1.3 Enumeration of Microstates and the Macrostate

The molecules that constitute a system are distributed over the energetic levels with N_j molecules on each level. The total number of molecules in the system is therefore $N = \sum_i N_i$, which can be distributed at any instance in a particular distribution over the energy levels, forming what is known as a *macrostate*. This distribution, called the *population distribution*, results in the total energy of the system, $E = \sum_i \varepsilon_i' N_i$. At a different time, because of energy exchange through collisions, the population distribution may change and another macrostate ensue. Over time, a certain macrostate will occur with greater frequency than the others. This is *the most probable macrostate* and will reflect the condition of thermodynamic equilibrium. Boltzmann's equation, $S = k \ln \Omega$, makes the connection between the molecular population distribution leading to the most probable macrostate, as the thermodynamic equilibrium condition, and the macroscopic thermodynamic property of entropy. Because each energy level can have a number of degeneracies, there can be a number of ways in which the molecules can be distributed over the energy levels without changing the overall macrostate energy. Thus, within the same macrostate, there can be a number of *microstates*. Figure 3.9 shows two population distributions in which the second energy level, ε_1', has two degeneracies with five molecules occupying these degeneracies in two different ways, therefore indicating two distinct microstates. However, the macrostate described by the overall energy of the system remains unchanged. The macrostate with the largest number of microstates is the one that appears with the most probability and describes, as indicated before, the system's thermodynamic equilibrium. Counting the number of microstates in a macrostate thus leads to the identification of the most probable macrostate and the link to the system's macroscopic properties.

3.4.2 Counting the Number of Microstates for a Given Macrostate

Finding the total number of microstates, Ω, is a two-step process: (i) finding the number of microstates for a given macrostate and (ii) summing all these numbers for all possible macrostates for the system. The number of microstates is determined by the number of ways in which N_i undistinguishable particles (molecules or atoms) can be distributed over the degeneracies in the available

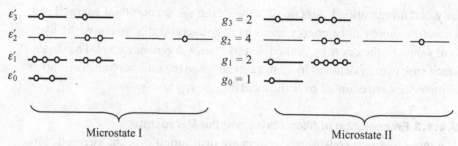

Figure 3.9. Population distribution for two selected microstates for a given macrostate.

states. In terms of the rules governing the distribution of the particles over the existing states, two statistical models exist: the *Bose–Einstein statistic* and the *Fermi–Dirac statistic*. Both are described in the subsections that follow.

BOSE–EINSTEIN STATISTIC. This group contains those molecules and atoms composed of an even number of elementary particles. For example, carbon ^{12}C has six protons and six neutrons, an even number of elementary particles, and therefore obeys the Bose–Einstein statistic. For this group there is no restriction on the number of particles in a state. The total number of microstates when only the indistinguishable distributions are considered is given by

$$\frac{[N_i + (g_i - 1)]!}{N_i!(g_i - 1)!}. \tag{3.36}$$

This expression applies to one energy level, and thus, when all possible microstates are summed over all energy levels, the total number of microstates for a given macrostate becomes

$$W(N_i) = \prod_i \frac{[N_i + (g_i - 1)]!}{N_i!(g_i - 1)!}. \tag{3.37}$$

$W(N_i)$ is known as the *thermodynamic probability* and is a measure of the degree of "disorder" of the system.

FERMI–DIRAC STATISTIC. This group includes the atoms and molecules with an uneven number of elementary particles. For example, atomic hydrogen has one proton and zero neutrons in its structure and obeys the Fermi–Dirac statistic. In this model, there can be no more than one particle in a degeneracy, which means that the number of particles is always smaller than the number of available degeneracies. Counting all the possible microstates for this statistic over all of the existing energy levels, we obtain the thermodynamic probability as

$$W(N_i) = \prod_i \frac{g_i!}{(g_i - N_i)! \, N_i!}. \tag{3.38}$$

3.4.3 The Most Probable State

The most probable macrostate contains the maximum number of microstates. For convenience, $\ln W$ is used instead of W, leading to,

$$\ln W = \sum_i [\ln(N_i + g_i - 1)! - \ln(g_i - 1)! - \ln N_i!], \tag{3.39}$$

$$\ln W = \sum_i [\ln g_i! - \ln(g_i - N_i)! - \ln N_i!] \tag{3.40}$$

for the Bose–Einstein and Fermi–Dirac statistics, respectively. For very large numbers of N_i and g_i, which is the case for the thermodynamic systems of interest here, $\ln W$ may be treated as a continuous function of N_i. Sterling's formula,

$$\ln x! \cong x \ln x - x, \tag{3.41}$$

can then be applied to expressions (3.36) and (3.37), which simplify to

$$\ln W = \sum_i \left[\pm g_i \ln \left(1 \pm \frac{N_i}{g_i} \right) + N_i \ln \left(\frac{g_i}{N_i} \pm 1 \right) \right], \tag{3.42}$$

where the $+$ sign applies to Bose–Einstein statistics and the $-$ sign applies to the Fermi–Dirac statistics. The maximum value of W is found from

$$d(\ln W) = 0 \tag{3.43}$$

under the constraints that $N = \sum_i N_i$ and $E = \sum_i \varepsilon_i' N_i$. We find solution for Eq. (3.40) by using Lagrange multipliers (for a complete derivation see Vincenti and Kruger, 1986), obtaining the particular values of N_i that maximize W:

$$\frac{N_i^*}{g_i} = \frac{1}{e^{\alpha + \beta \varepsilon_i'} \mp 1}, \tag{3.44}$$

where α and β are the Lagrange multipliers to be found from the constraints $N = \sum_i N_i$ and $E = \sum_i \varepsilon_i' N_i$. The asterisk indicates that these are the particular values of N_i that give the most probable macrostate. Again, the $-$ sign is for the Bose–Einstein statistic and the $+$ sign is for the Fermi–Dirac statistic.

3.4.4 The Boltzmann Distribution

At low temperatures of several degrees Kelvin, the particles are concentrated in the ground level, but at elevated temperatures the particles are distributed over a large number of energy levels; therefore the energy levels are sparsely populated, which means that $N_i \sqrt{g_i}$. Hence $e^{\alpha + \beta \varepsilon_i'} \, ^{TM} \, 1$ and Eq. (3.40) simplify to

$$N_i^* = g_i \, e^{-\alpha} e^{-\beta \varepsilon_i'}. \tag{3.45}$$

This limiting case is known as the Boltzmann limit, and Eq. (3.45) is called the Boltzmann distribution. The Lagrange multiplier α is eliminated from the constraints, and β is found from the link between classical and statistical thermodynamics as $\beta = 1/kT$, where k is the Boltzmann constant and T is the system temperature (Vincenti and Kruger, 1986). Equation (3.45) then becomes

$$N_i^* = N \frac{g_i e^{-\varepsilon_i'/kT}}{\sum_i g_i e^{-\varepsilon_i'/kT}}. \tag{3.46}$$

The quantity in the denominator is called the *partition function*:

$$Q \equiv \sum_i g_i e^{-\varepsilon_i'/kT}. \tag{3.47}$$

Equation (3.46) makes it evident that each term in the sum is proportional to the number of particles in the group of energy states; in other words, it emphasizes how the particles are partitioned among the energy groups.

3.4.5 Thermodynamic Properties in Terms of the Partition Function

The link between statistical thermodynamics and macroscopic properties of a system is expressed in Boltzmann's equation through the association of the entropy with the degree of disorder of the macroscopic system and the interpretation of the total number of microstates as a measure of system disorder from the statistical thermodynamic point of view. Because the partition function is shown to equal the maximum number of microstates for a system composed of a large number of particles (Vincenti and Kruger, 1986), the Boltzmann equation serves as a basis to calculate a system's macroscopic properties from knowledge of its microscopic properties.

In the Boltzmann limit, Eq. (3.42) becomes

$$\ln \Omega = \sum_i N_i^* \left(\ln \frac{g_i}{N_i^*} + 1 \right), \tag{3.48}$$

which, along with Eq. (3.46) and the definition of partition function (3.47), when substituted into the Boltzmann equation, results in

$$S = kN \left(\ln \frac{Q}{N} + 1 \right) + \frac{E}{T}. \tag{3.49}$$

The internal energy E is obtained from the equilibrium constraint $E = \sum_i \varepsilon_i' N_i^*$, in which the total number of particles in the most probable macrostate, N_i^*, is substituted from Boltzmann distribution expression (3.46) to obtain

$$E = NkT^2 \left(\frac{\partial \ln Q}{\partial T} \right)_v, \tag{3.50}$$

where the partition function has been differentiated with respect to temperature at constant volume. With the observation that $Nk/m = R$, the specific-gas constant, the internal energy per unit mass is

$$e = RT^2 \left(\frac{\partial \ln Q}{\partial T} \right)_v. \tag{3.51}$$

Combining Eqs. (3.49) and (3.51) we obtain an expression for the entropy in terms of the partition function:

$$S = kN \left(\ln \frac{Q}{N} + 1 \right) + NkT \left(\frac{\partial \ln Q}{\partial T} \right)_v. \tag{3.52}$$

Additional thermodynamic properties can be obtained as functions of the partition function. The first law of classical thermodynamics in differential form is

$$T \left(\frac{\partial S}{\partial V} \right)_T = \left(\frac{\partial E}{\partial V} \right)_T + pdV, \tag{3.53}$$

which provides an expression for pressure:

$$p = NkT \left(\frac{\partial \ln Q}{\partial V} \right)_T. \tag{3.54}$$

The macroscopic thermodynamic properties e, S, and p given by Eqs. (3.51), (3.52), and (3.54) can now be determined if the appropriate expression for the partition function can be found as a function of temperature and volume.

3.4.6 Evaluation of the Partition Function

In the definition of the partition function, $Q \equiv \sum_i g_i e^{-\varepsilon_i/kT}$, the energy levels and their degeneracies are needed. The quantized levels for translation, rotation, vibration, and electronic energies were determined by quantum mechanics and are listed for a number of species in the literature (Herzberg, 1989; NIST-JANNAF, 1998). The total energy of the molecules that appears in the exponent in the partition function expression is formed from the translational, rotational, vibrational, and electronic contributions:

$$\varepsilon' = \varepsilon'_{\text{trans}} + \varepsilon'_{\text{rot}} + \varepsilon'_{\text{vib}} + \varepsilon'_{\text{el}}. \tag{3.55}$$

Substituting into the partition function results in an expression that includes a product of mathematical expressions that each depend on a different energy mode. Thus the partition function can be written as a product of various internal partition functions:

$$Q = Q_{\text{trans}} Q_{\text{rot}} Q_{\text{vib}} Q_{\text{el}}. \tag{3.56}$$

It remains to evaluate each of the participating functions for the given thermodynamic system. For monoatomic gases only the translational and the electronic partition functions are necessary, but for diatomic and polyatomic gases the internal structure contributing to rotation and vibration are needed as well. Without derivation (given, for example, in Vincenti and Kruger, 1986, and Herzberg, 1989), the partition functions take the following forms:

- For translation,

$$Q_{\text{trans}} = V \sqrt{\left(\frac{2\pi m k T}{h^2}\right)^3},$$ (3.57)

  where h is Planck's constant.
- For rotation, with the assumption of temperature in excess of several degrees Kelvin, the partition function of diatomic gases is

$$Q_{\text{rot}} = \frac{T}{\Theta_r},$$ (3.58)

where Y_r is the characteristic temperature for rotation given by $\Theta_r \equiv h^2/8\pi^2 I k$, where I is the molecule's moment of inertia. Polyatomic molecules with a linear geometry have a similar partition function for rotation with the modification that $Q_{\text{rot polyatomic}} = T/\sigma\Theta_r$, where $\sigma = 1$ for molecules without a center of symmetry and $\sigma = 2$ for molecules with a center of symmetry, e.g., CO_2. Nonlinear polyatomic molecules have three distinct moments of inertia, and their partition function for rotation is

$$Q_{\text{rot polyatomic}} = \frac{1}{\sigma} \sqrt{\frac{\pi}{ABC} \left(\frac{kT}{h}\right)^3},$$ (3.59)

where A, B, and C are modified principal moments of inertia, with $A = h/(8\pi^2 I_A)$, and so on, and σ is the number of indistinguishable molecule orientations.

- The vibration partition function based on the model of a simple harmonic oscillator results in the following expression:

$$Q_{\text{vib}} = \prod \frac{1}{\left(1 - e^{-h\nu_i/kT}\right)^{g_i}},$$ (3.60)

  where ν_i are the molecule's vibrational frequencies and g_i are the degeneracies associated with these frequencies.
- The electronic partition function derives directly from its definition,

$$Q_{\text{el}} \equiv \sum_i g_i e^{-\varepsilon_i'/kT},$$ (3.61)

with the observation that in most cases the first few terms dominate and the others can be reasonably neglected.

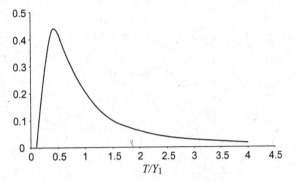

Figure 3.10. Electronic specific-heat dependence on temperature.

3.4.7 Evaluation of Thermodynamic Properties

Once the partition functions are determined, the thermodynamic properties result from the relations established in the previous analysis. The internal energy associated with translation is obtained from Eq. (3.51),

$$e_{\text{trans}} = \frac{3}{2} RT, \tag{3.62}$$

which is agreement with the kinetic theory of *equipartion of energy*, which states that each thermal degree of freedom contributes $1/2\,RT$ to the energy per unit mass. The specific heat at constant volume for a monoatomic gas, $c_v \equiv (\partial e / \partial T)_v$, is

$$c_v = \frac{3}{2} R. \tag{3.63}$$

Assuming that only the first two terms in the electronic partition function are sufficiently large, the electronic contribution to the internal energy is

$$e_{\text{el}} = R\Theta_1 \frac{(g_1/g_0)\, e^{-\Theta_1/T}}{1 + (g_1/g_0)\, e^{-\Theta_1/T}} \tag{3.64}$$

where $\Theta_1 = \varepsilon_1'/k$ is the characteristic temperature for the second term in the partition function, and the specific heat is

$$c_{v_{\text{el}}} = R\left(\frac{\Theta_1}{T}\right)^2 \frac{(g_1/g_0)\, e^{-\Theta_1/T}}{[1 + (g_1/g_0)\, e^{-\Theta_1/T}]^2} \tag{3.65}$$

The variation of $c_{v_{\text{el}}}/R$ as a function of T/Θ_1 is shown in Fig. 3.10, indicating that a maximum exists around $T/\Theta_1 \approx 0.4$. From the spectroscopic determination of the electronic energy levels it was found that, for a number of species, the higher terms are reducing in fact to constants that disappear in the differentiation of the partition function and the maximum in the electronic specific-heat curve occurs at temperatures much lower than those at which

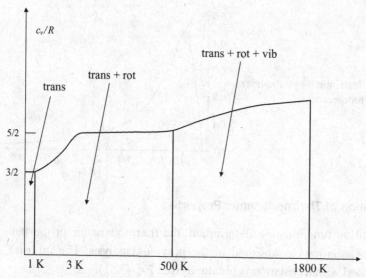

Figure 3.11. Specific-heat variation with temperature for a diatomic gas.

the monoatomic species are usually formed. For example, for atomic oxygen the maximum occurs around 100 K, whereas dissociation of molecular oxygen begins above 1800 K, as shown in Fig. 3.11. Therefore, in most cases, the electronic contribution to the internal energy and the specific heat is negligible.

For rotation, the internal energy and the specific heat are obtained as

$$e_{\text{rot}} = RT, \quad c_{v_{\text{rot}}} = R. \tag{3.66}$$

For vibration, the internal energy is obtained as

$$e_{\text{vib}} = \frac{R\Theta_v}{e^{\Theta_v/T}-1} \tag{3.67}$$

and the vibrational specific heat as

$$c_{v_{\text{vib}}} = R\left[\frac{\Theta_v/2T}{\sinh\left(\Theta_v/2T\right)}\right]^2. \tag{3.68}$$

The total specific heat for diatomic molecules is thus found as a sum of all the contributions:

$$c_v = \frac{5}{2}R + R\left[\frac{\Theta_v/2T}{\sinh\left(\Theta_v/2T\right)}\right]^2 + c_{v_{\text{el}}}, \tag{3.69}$$

where the electronic specific heat has been maintained for completeness, although, as noted, it is often neglected in practice. This expression of the specific heat indicates, as shown in the diagram in Fig. 3.11, that, for a range of moderate temperatures, before the vibrational excitation becomes significant, the specific heat of a diatomic will remain essentially constant except for the region of temperatures close to absolute zero when the rotational motion essentially stops and the gas maintains only translational temperatures. At elevated temperatures, the vibrational contribution becomes significant and the specific-heat value increases until dissociation changes the gas chemical composition.

An expression for the system pressure is obtained from Eq. (3.54) with the observation that only the translational partition function depends on volume. Hence, using Eqs. (3.57) and (3.54), we obtain the pressure as

$$p = \frac{NkT}{V} \quad \text{or} \quad p = \frac{RT}{V} \tag{3.70}$$

when derived for 1 mol of gas. Thus the equation of state for an ideal gas is regained, in which the assumptions of independent and indistinguishable particles implicitly indicate the absence of intermolecular forces.

3.5 Hypersonic Flow

The hypersonic definition of an "ultra high" supersonic regime is attributed to H. S. Tsien (Cummings and Yang, 2003), who presented his work around the same time von Karman made his approximations for transonic flow. Therefore Tsien too used the velocity potential to derive similarity laws for the hypersonic flow. Identifying the difficulties imposed in the theoretical treatment of hypersonic flow, von Karman noted in 1955 that the disturbance velocity, although small in comparison with the velocity component, may not be small in comparison with the speed of sound, and, with this observation, similarity rules can be derived for hypersonic flow in a similar fashion as is done traditionally for transonic flows.

The jump conditions given by Eqs. (2.28) for normal flow, under assumptions of ideal flow, are extended to oblique shocks with the inclusion of the angle β, made by the velocity vector with the oblique shock. In the limit of large Mach numbers, the traditional jump equations for oblique shocks (Liepmann and Roshko, 1957) reduce to (Anderson, 1989)

$$\frac{p_2}{p_1} = \frac{2\gamma}{\gamma + 1} \left(M_1^2 \sin^2 \beta - 1 \right),$$

$$\frac{T_2}{T_1} = \frac{2\gamma(\gamma - 1)}{(\gamma + 1)^2} \left(M_1^2 \sin^2 \beta - 1 \right),$$

$$\frac{\rho_2}{\rho_1} = \frac{\gamma + 1}{\gamma - 1}. \tag{3.71}$$

This set of jump equations can be complemented with the equation that relates the shock angle β to the deflection angle θ and Mach number M_1 (Liepmann and Roshko, 1957):

$$\tan \theta = 2 \cot \beta \left[\frac{M_1^2 \sin^2 \beta - 1}{M_1^2 (\gamma + \cos 2\beta) + 2} \right]. \tag{3.72}$$

It should be noted that the jump equations and (2.28) were derived for calorically perfect gases, an assumption that begins to fail at high speeds. The departure from this simplification is dictated by several physical phenomena, including

- the distribution of the thermal energy over the molecules' internal modes of energy,
- changes in the isentropic compression that are due to high temperatures,
- dissociation and ionization that may occur at elevated temperatures,
- viscous and heat transfer effects that become significant.

Despite these simplifications, an analysis of Eq. (3.72) offers interesting insights into the nature of hypersonic shock waves with the observation that the shock angle β becomes independent of the Mach number and, because the shock and the wall angles are small, $\sin \beta \approx \beta$, and the shock angle becomes proportional to the wall angle,

$$\beta \approx \frac{\gamma + 1}{2} \theta. \tag{3.73}$$

The proportionality constant is close to unity, particularly as the temperature increases at higher Mach numbers and the internal modes of energy are activated. The wave angle then becomes only slightly larger than the wedge that has generated it, and hence the layer between the shock wave and the body, defined as the *shock layer* (Anderson, 1989), is very thin. At the leading edge of the body, even the slightest bluntness generates a curvature of the shock, and thus there is a strong jump in the thermodynamic parameters across the shock in the region immediately in front of the leading edge. This large gradient in properties leads to a substantial entropy change within a thin layer that propagates along the wall for a long distance and is known as the *entropy layer*. Eventually the entropy layer will be ingested by the engine inlet capture, leading to flow distortions.

Von Karman (1955) made the observation that Newton's sine-squared law for a pressure coefficient would apply in the hypersonic approximation, with the observation that flow deflection occurs at the shock rather than at the wall. Anderson (1989) points out the close results obtained for the pressure coefficient through the Newtonian derivation compared with the derivation based on the jump conditions under the assumption of large Mach numbers when the wall angle is replaced with the shock angle:

$$c_p = 2\sin^2\theta, \quad \text{Newtonian sine-squared law,}$$

$$c_p \to 2\sin^2\beta, \quad \text{jump conditions with } M_1 \to \infty \text{ and } \gamma \to 1. \quad (3.74)$$

To complete this brief discussion about hypersonic effects, it should be noted that the product $M_1\sin\beta \approx M_1\beta$, because of the proportionality between the shock and the wall angle given by proportion (3.73), leads to the definition of the *hypersonic similarity parameter*, $K \equiv M_1\theta$. The jump conditions given in Eqs. (3.71) can now be expressed as functions of the hypersonic similarity parameter K.

REFERENCES

Anderson, J. D. (1989). *Hypersonic and High Temperature Gas Dynamics*, McGraw-Hill.

Chapman, S. and Cowling, T. G. (1952). *The Mathematical Theory of Non-uniform Gases*, 2nd ed., Cambridge University Press.

Cullen, H. B. (1985). *Thermodynamics and an Introduction to Thermostatistics*, 2nd ed., Wiley.

Cummings, R. M. and Yang, H.-T. (2003). "Lester Lees and hypersonic aerodynamics," *J. Propul. Power* **40**, 467–474.

Eckert, E. R. G. and Drake, Jr., R. M. (1959). *Heat and Mass Transfer*, 2nd ed., McGraw-Hill Series in Mechanical Engineering.

Emanuel, G. (1987). *Advanced Classical Thermodynamics* (J. S. Przemieniecki, ed.), AIAA Educational Series.

Herzberg, G. (1989). *Molecular Spectra and Molecular Structure*, Vols. I and II, reprint of the 2nd ed., Krieger.

Hirshfelder, J. O., Curtiss, C. F., and Bird, R. B. (1954). *Molecular Theory of Gases and Liquids*, Wiley.

Liepmann, H. W. and Roshko, A. (1957). *Elements of Gasdynamics*, 2nd ed., Wiley.

Mason, E. A. (1969). "Transport in neutral gases," in *Kinetic Processes in Gases and Plasmas* (A. R. Hochstim, ed.), Reentry Physics Series, Academic Press.

NIST-JANNAF Thermochemical Tables. (1998). *J. Phys. Chem. Ref. Data*, 4th ed.

Vincenti, W. G. and Kruger, C. H. (1986). *Introduction to Physical Gas Dynamics*, Krieger.

Von Karman, T. (1955). "Solved and unsolved problems in high speed flows," in *Proceedings of the Conference on High-Speed Aerodynamics* (A. Ferri, N. J. Hoff, and P. A. Libby, eds.), Polytechnical Institute of Brooklyn.

4 Cycle Analyses and Energy Management

4.1 Introduction

Given the broad range of aerothermodynamic conditions experienced during hypersonic flight, the most likely engine design would include some combination of a scramjet operation with other types of propulsion systems. Examples of combinations of propulsion systems or of combined cycles are numerous, and several were described in Chap. 1. As speed increases, the integration between vehicle aerodynamics and engine performance becomes more and more coupled; the vehicle forebody becomes part of the engine intake and the vehicle aft becomes part of the nozzle; engine throttling changes the pressure distribution to a degree that substantially modifies the moments acting on the vehicle. These vehicle–engine-coupling considerations include not only the flow field generated on the vehicle forebody and afterbody but also structural cooling requirements that, in turn, become increasingly higher with the flight speed; they must be satisfied by the fuel on board. This close coupling requires that, in fact, the selection of the engine cycle be dictated by the entire system optimization.

The differences between the ramjet, in which combustion takes place at subsonic speeds, and the scramjet, which maintains supersonic speeds throughout the entire engine, were outlined in Section 1.2. Performance-based differences between different engine cycles are clearly illustrated in the specific impulse of the fuel, $I_{sp} = \frac{thrust}{gravimetric\,fuel\,rate}$, whose diagram is shown in Fig. 4.1 (after Billig, 1996a). Here, the specific impulse is shown in relative units to eliminate the dependence on the fuel used. This diagram shows that around Mach 3 the subsonic combustion ramjet becomes more efficient as a propulsive system in comparison with the turbine-based engines (turbojets of turbofans), but beyond Mach 5 its performance begins to decay rapidly and the scramjet delivers a higher specific impulse. The rocket's specific impulse is considerably lower than that of the other propulsion system, but it is included in this diagram because it is the only system that offers operational capabilities

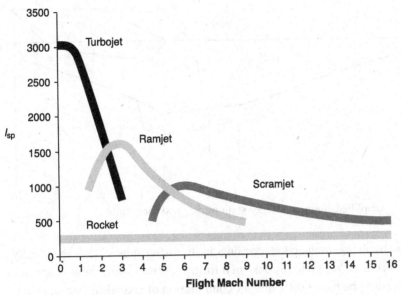

Figure 4.1. Specific impulse as a function of the flight Mach number for selected engine cycles.

from sea-level static to beyond the atmosphere. The rocket's low specific impulse in comparison with that of the other propulsion systems clearly eliminates it from consideration for long-range cruises; however, as the Mach number continues to increase in the hypersonic regime, the scramjet's specific impulse approaches that of the rocket engine and, combined with the continually decreasing air density, the engine will have to transition to rocket operation for orbital flight. Historically, multiple-staged vehicles have been designed to operate with a single type of propulsion system for each stage. Stages are optimized for different altitude and Mach number regimes in the trajectory increasing the overall specific impulse of the system. As a recent example, NASA's hypersonic aircraft demonstrator, X-43, uses subsonic aircraft propulsion as the first stage followed by a second stage provided by a rocket, Pegasus, for supersonic acceleration and the scramjet-based research vehicle as the third stage. The limited-operation scramjet-powered vehicle began its autonomous flight at M > 7.

Clearly the optimization for long-range, broad-speed-regime operation requires the use of a combination of some of these propulsion systems, which, when operated in a synergistic way, achieve performance enhancement over individual cycles; in this case they are referred to as combined-cycle propulsion (CCP) systems. These systems, discussed in more detail in Section 4.5, can be broadly divided into two categories: combined cycles that could include turbojet or turbofan cycles and systems that combine the scramjet operation with rockets. Air-breathing combined-cycle engines are particularly useful for

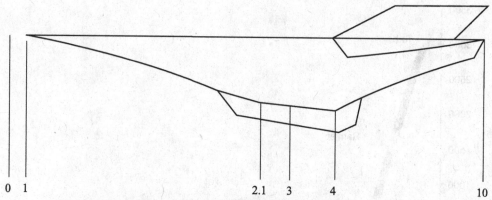

Figure 4.2. Simplified engine reference stations. See text for details.

missions involving high-speed cruising in the atmosphere and eventually reaching orbit. Other systems, designed for atmospheric operation of shorter duration, could be based on a simpler combination of propulsion systems that operate in sequence rather than in synergy.

The following sections include generalized discussions of scramjet cycle analysis and performance and integration of the scramjet engine with the vehicle and in combination with other propulsion systems. Component analyses and integration are discussed in Chaps. 5 and 6.

4.2 Ideal Scramjet Cycle

The scramjet engine belongs to the family of Brayton cycles, which consist of two adiabatic and two constant-pressure processes (for detailed analyses see, for example, Mattingly et al., 2002). A simplified schematic of a scramjet-equipped vehicle is shown in Fig. 4.2, assuming a lifting body with the vehicle's forebody performing a large part of the inlet compression; the afterbody constitutes part of the nozzle. The engine therefore occupies the entire lower surface of the vehicle. The standard engine designation, which was adopted here after Heiser and Pratt (1994), derives from the standard station designations of gas-turbine engines and is used to emphasize the separation between the major engine components:

- Station 0 represents the free-stream condition.
- Station 1 represents the beginning of the compression process. Hypersonic shock-wave angles are small, resulting in long compression ramps (or spikes if an axisymmetric configuration is used) that, in many of the suggested configurations, begin at the vehicle's leading edge. Additional compression takes place inside the inlet duct.

- Station 2.1 represents the entrance into the isolator section. The role of the isolator is to separate the inlet from the adverse effects of a pressure rise that is due to combustion in the combustion chamber. The presence of a shock train in the isolator further compresses the air before arriving at the combustion chamber. Thermodynamically the isolator is not a desirable component, because it is a source of additional pressure losses, increases the engine cooling loads, and adds to the engine weight. However, operationally it is needed to include a shock train that adjusts such that it fulfills the role just described.
- Station 3 is the combustion chamber entrance. Unlike the turbojet engine cycle, in which the air compression ratio is controlled by the compressor settings, in a fixed-geometry scramjet the pressure at the combustion chamber entrance varies over a large range.
- Station 4 is the combustion chamber exit and the beginning of expansion.
- Station 10 is the exit from the nozzle; because of the large expansion ratios, the entire aft part of the vehicle may be part of the engine nozzle.

Clearly, this selection of engine stations represents an idealized component separation that omits the presence of additional elements – for example, the possible presence of a magnetohydrodynamic (MHD) generator in the inlet's diffuser coupled with a MHD accelerator in the nozzle's expansion section (Burakhanov et al., 2001) – that may be included in the engine design or the interaction with other propulsion cycles that, together with the scramjet, constitute a combined-cycle engine.

The engine components' efficiency plays an increasingly significant role as the kinetic energy of the airstream increases. As the flight Mach number increases, the kinetic energy of the air far exceeds the heat released through combustion, and thus the net thrust becomes only a small fraction of the airstream thrust entering the engine. For example, Anderson et al. (2000) estimated that when flying at Mach 16, using hydrogen in stoichiometric proportions, the energy added through combustion is only one fourth that of the airstream's kinetic energy. Kinetic energy management and its implications for component efficiency become critical issues for optimization under the given constraints.

The idealized engine cycle can be easily analyzed by use of the entropic diagram shown in Fig. 4.3. As is the general practice, it is assumed that the air captured by the inlet remains of the same composition throughout the engine and that the combustion process is replaced with heat addition. The additional mass introduced by the fuel is small compared with the air mass flow and can therefore be neglected without introducing significant errors in the cycle analysis. Finally, a constant-pressure transformation is introduced to close the cycle

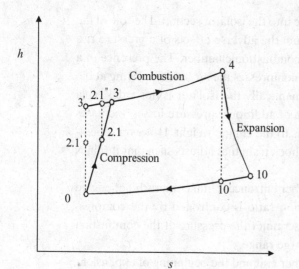

Figure 4.3. Ideal scramjet cycle. The initial and final thermodynamic states in terms of static specific enthalpy and static specific entropy for each component are indicated by the engine station numbers.

and return to the original thermodynamic state. These processes are subsequently described in some detail.

The compression process from Station 0 to Station 2.1 is achieved by flow deceleration through a system of shock waves generated on the forebody upstream of the inlet capture through what is referred to as external compression and continues inside the duct through internal compression. Along the forebody, a substantially thick boundary layer forms, and, because the engine cannot be displaced from the vehicle to avoid boundary-layer ingestion and the negative effects of shock-wave–boundary-layer interactions that can lead to separation, some of the incoming boundary layer will be removed through bleed. These quantities of air are, however, small, and their effect on the thermodynamic evolution during compression can be neglected in the entropic diagram.

The degree of flow deceleration in the inlet is dictated by the constraints of velocity and static temperature at the combustion chamber entrance, while ensuring (i) high efficiency and (ii) matching with additional flow streams that may be present in a combined-cycle engine. The inefficiencies of the compression process, which appear in the entropic diagram as a departure from the isentropic compression, depend on the kinetic energy transformation and its level at the end of the compression process and have obvious implications for the efficiency of the other engine components and the system as a whole. Most notably, the static temperature rise that is due to inefficiencies in the inlet may lead to dissociations that reduce the heat released during the combustion process. Finally, the compression process is influenced by the interactive processes resulting from complex, 3D fluid dynamic interactions that include multiangled shock waves, interaction between shock waves and boundary layers, separation vortices, and vortex–vortex interactions. Certain designs suggested the addition of fuel during compression in the inlet to increase the residence time for improved mixing (Guoskov et al., 2001; Vinogradov et al., 2001), thereby introducing mass, changing the properties of the incoming air, and affecting the inlet kinetic efficiency. The level of flow distortions generated in the inlet, both steady state and dynamic, has a clear effect on the heat-release processes in the combustion chamber through distortion interaction with mixing and turbulence–temperature effects (Warnatz et al., 1996; Oran and Boris, 2001). Inlet-flow distortion effects on the compressors in turbojet engines were studied and documented extensively (Younghans, 1989); however, the effect of inlet distortions in ramjet–scramjet engines has not been evaluated sufficiently to date.

The process in the isolator between Stations 2.1 and 3 can be considered as part of the compression process, although the isolator has a clearly defined function: to protect the inlet flow from the pressure changes in the combustion chamber. This compression is the result of the shock train present in the isolator that, depending on the flight regime, may extend in the core of the combustion chamber surrounded by regions of subsonic flow or end with a normal shock, thus rendering the entire flow subsonic before arriving at the combustion chamber. The inefficiencies in the isolator result from viscous losses, heat lost to the walls, and shock–boundary-layer interactions.

Figure 4.3 includes the idealized isentropic compression in the inlet between states 0 and 2.1′ and the isentropic isolator compression between states 2.1 and 2.1″.

Between Stations 3 and 4 heat is released through fuel combustion. Assuming in a first approximation (Heiser and Pratt, 1994) that the enthalpy remains constant between the free stream and the entrance to the combustion chamber, i.e., Station 3, a relation between M_3 and M_0 can be written, which, in the limit of large flight velocities, of the order of $M_0 = 10$, becomes

$$\frac{M_3}{M_0} \approx \sqrt{\frac{T_0}{T_3}}. \tag{4.1}$$

Because the expected temperature ratios T_3/T_0 are of the order of 10, approximation (4.1) implies that the compression in the inlet–isolator results in M_3, which is approximately one third of M_0.

The departure from constant stagnation pressure in the combustion chamber is due mostly to friction, Rayleigh losses, and heat transferred to the wall (Heiser and Pratt, 1994). The amount of heat released within the combustion chamber depends on the efficiency of the mixing process and the degree of conversion of the available chemical energy into sensible energy. Following some sort of flameholding scheme, the combustion chamber is usually designed with a constant area for rapid heat release followed by a slowly expanding region to delay thermal-choking onset, which is particularly severe at low-speed operation as indicated by Eq. (2.34) and Fig. 2.2. At high speeds, this slowly diverging section of the combustion chamber acts as an initial expansion region during which the flow has additional time to reach chemical equilibrium (Ortwerth, 2000).

Expansion follows between Stations 4 and 10, first in an internal nozzle, and continues on the vehicle aft to, ideally, perfect expansion. The irreversibilities in the nozzle are caused by friction, viscous dissipation in shocks, and heat lost to the structure. If rapid expansion begins before chemical equilibrium has been achieved at the exit from the combustion chamber, a certain amount of dissociated species may freeze, leading to additional energy loss. The degree of expansion results from optimization of engine performance, vehicle dimensions, and requirements of balancing the moment for trimmed flight.

Finally, the cycle is completed with the imaginary process from 10 to 0, which represents heat rejection at constant pressure, equivalent to the difference between the thermodynamic conditions at the nozzle exit and at the free stream.

This is a simplified representation of the scramjet thermodynamic cycle. It is expected that the scramjet will operate with other propulsion systems in a combined cycle so that the propulsion system–vehicle can be optimized for the entire flight regime. The corresponding thermodynamic cycle will differ from the ideal cycle presented here, according to the engine configuration selected. Some of these cycles are reviewed in Section 4.7. A discussion of scramjet components' thermodynamic efficiencies and constraints is included in Chaps. 5 and 6.

4.3 Trajectory and Loads

Optimization of the scramjet-powered vehicle trajectory takes into account the mission requirements, such as insertion into low Earth orbit (LEO) following air-breathing propulsion for a transatmospheric flight (Hargraves and Paris, 1987) or a more restrictive mission for hypersonic missiles (Bowcutt, 2001), within the constraints dictated by size, structural loads, and operational features, such as the transition from the initial propulsion cycles to scramjet and then to rocket propulsion.

Possibly the most critical mission parameter is the maximum payload. On the whole, it can be maximized by the minimization of fuel consumption. Therefore an optimal trajectory may be inferred from energy–altitude analyses (Bryson et al., 1969; Schmidt, 1997) in which global minimization of fuel consumption results from a maximization of the energy level with respect to fuel consumption, dE/dW_{fuel}. Under the energy-state assumption (Schmidt and Hermann, 1998), the energy change with respect to fuel consumption can be related to the flight conditions, i.e., thrust T, drag D, and specific impulse I_{sp}:

$$\frac{dE}{dW_{fuel}} \simeq \frac{V I_{sp}}{W} \left(1 - \frac{W}{T}\frac{D}{L}\right), \tag{4.2}$$

and contours of constant dE/dW_{fuel} can be sketched on an altitude–velocity map. The loci of the curves' maxima describe an optimal trajectory with respect to the minimum fuel consumption that has been selected as the optimization parameter in this case.

Similarly, the trajectory can be optimized with respect to the optimal time to reach a desired altitude, with the time to change the altitude obtained from

$$\Delta t = \int_{E_1}^{E_2} \frac{dE}{dE/dt} = \int_{E_1}^{E_2} \frac{W}{(T_t - D)V} dE. \tag{4.3}$$

The trajectory optimized for one of these two conditions can be maintained as long as the constraints in the system are satisfied, including (i) energy addition through combustion and (ii) the dynamic pressure limit.

On the same basis of the kinetic energy of the air ingested by the scramjet engine and under the constraint of energy availability from the fuel, Czysz and Murthy (1995) separated the scramjet engine operation into five regimes, depending on flight velocity. These regimes reflect, in order of increasing flight velocity, the ability to add energy to the air through combustion that works in competition with the engine drag losses. At moderate hypersonic flight velocities, a significant amount of heat can be added to the airflow while the engine drag losses are relatively moderate, resulting in high acceleration capabilities. The situation gradually changes as relative heat addition to the air progressively decreases with increased flight velocity whereas the drag losses (Riggins, 1997; Mitani et al., 2002) continuously increase until the heat addition can no longer overcome the drag and the air-breathing-based system reaches the extent of its flight envelope.

Optimizations of vehicle architecture under the constraints of thermal and mechanical loads, external drag, and internal irreversibilities generally indicated in recent studies (Trefny, 1999; Mehta and Bowles, 2001) that trajectories for both SSTO and TSTO concepts are defined by a flight dynamic pressure residing between 4500 and 9000 kg/m^2 (about 1000–2000 lb/ft^2). The lower values are needed to reduce the aerodynamic drag during the transonic flight transition, and the higher values are recommended for the hypersonic acceleration.

A typical trajectory corridor is shown in Fig. 4.4, which includes the range of altitude Mach numbers experienced during the flights with the experimental Kholod vehicle in the early 1990s (Semenov et al., 2002). Higher dynamic pressure ratios are expected during reentry from orbital flight and

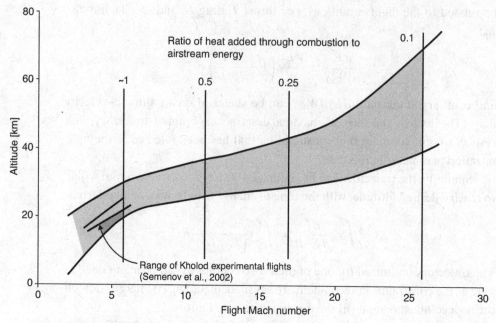

Figure 4.4. Flight regimes indicating the ability to add energy through combustion and the limit of thrust production (after Czysz and Murthy, 1995).

deceleration when drag is used to dissipate the kinetic energy. Also included in this figure are the estimates of energy added by combustion as a percentage of the airstream energy as the flight Mach number increases (Czysz and Murthy, 1995), which indicate that around Mach 25 the scramjet engine reaches an energetic envelope limit.

4.4 Performance Analysis

In the Breguet range equation,

$$R = V I_{\mathrm{sp}} \left(L/D\right) \ln\left(W_0/W_f\right), \tag{4.4}$$

where W_0/W_f is the ratio of the vehicle weights at the beginning and the end of the cruise segment, V is the flight speed, and the $V I_{\mathrm{sp}}$ term is essentially constant for a scramjet engine in the domain of a hypersonic flight regime (Ortwerth, 2000). The vehicle will thus be required to fly at the best L/D and the product $V I_{\mathrm{sp}}$ will have to be maximized. This parameter depends on the propulsive efficiency, the efficiency of the chemical-to-thermal-energy conversion in the combustion chamber, and the efficiency of the other thermodynamic processes in the engine components.

Through the analysis of the scramjet cycle shown in Fig. 4.3, the nozzle exit velocity is obtained as

$$V_{10}^2 = V^2 + 2H_0(\psi - 1)\left[\eta_c\eta_e\left(\frac{\Delta H_c/H_0}{\psi}\right) - 1\right], \tag{4.5}$$

where H_0 is the flight enthalpy, ΔH_c is the energy added through chemical reactions, and η_c and η_e are the inlet compression and nozzle expansion thermodynamic efficiencies, respectively. These two efficiencies are defined as the departure of the compression and the expansion static enthalpy changes during these processes from the equivalent adiabatic evolution between the same isobars, i.e.,

$$\eta_c = \frac{h_{3'} - h_0}{h_3 - h_0}, \quad \eta_e = \frac{h_4 - h_{10}}{h_4 - h_{10'}}. \tag{4.6}$$

The parameter ψ in the specific-impulse equation represents the inlet static enthalpy rise through compression, $\psi = H_3/H_0$, and is limited by the maximum allowable compression temperature determined from considerations of dissociation discussed in Section 4.1. Thus the specific impulse defined as the thrust-to-mass-flow ratio becomes

$$I_{sp} = \frac{F}{\dot{m}g} = \frac{1}{g}(V_{10} - V)$$

$$= \frac{V}{g} \left\{ \sqrt{1 + \frac{2}{(\gamma - 1)M_0^2}(\psi - 1)\left[\eta_c \eta_e\left(\frac{\Delta H_c/H_0}{\psi}\right) - 1\right]} - 1 \right\}, \tag{4.7}$$

where H_0/V^2 has been simplified under the assumption that the air behaves as a thermally perfect gas and the flight Mach number has been emphasized.

Notable in this equation are the compression and the expansion efficiencies. Implicitly, the specific-impulse equation contains the combustion efficiency defined as

$$\eta_b = \frac{\Delta H_c}{f H_f} = \frac{H_4 - H_3}{f H_f}, \tag{4.8}$$

where f is the fuel-to-air mass-flow ratio and H_f is the fuel-heating capacity through the term ΔH_c. The combustion efficiency depends on the geometric configuration of the combustion chamber and the air and fuel thermodynamic properties that enter the combustion chamber and is closely coupled with the efficiency of the mixing process. The combustion efficiency is discussed in more detail in Chap. 6. Thus Eqs. (4.6) emphasize the effects of the component efficiencies on the specific impulse, which, as shown in Section 4.3, is a key parameter in vehicle trajectory and sizing optimization.

It would appear from Eq. (4.7) that the efficiencies of these three components (the inlet–isolator group η_c, the combustion chamber η_b, and the nozzle η_e) have the same quantitative effect on the specific impulse. Yet a lower compression efficiency leads to an increased temperature T_3, which reduces the amount of heat that can be released in the combustion chamber, thereby changing the mixing and combustion efficiencies in the combustion chamber and, by modifying the thermodynamic properties at the combustion chamber exit, influencing the expansion in the nozzle and the efficiency of this process.

For this reason, a thermodynamic cycle analysis based on an individual component sequence provides only qualitative results and is, in principle, similar to other Brayton cycle analyses.

From the point of view of a cycle thermodynamic analysis, it is interesting to point out other global parameters of engine efficiency, in particular the propulsive efficiency, which is defined as the engine exit energy realized out of the total energy available (Bullock, 1989). With the assumption that there is no mass change throughout the engine, the propulsive efficiency becomes

$$\eta_p = \frac{2V_0}{V_{10} + V_0}. \tag{4.9}$$

This parameter indicates that the maximum propulsive efficiency is obtained when the nozzle exit velocity equals the flight velocity, a condition in which, evidently, no specific impulse is created.

The propulsive efficiency is often combined with a thermal efficiency, η_{th}, which is defined as the kinetic energy increment across the entire engine normalized by the amount of energy contained in the fuel consumed:

$$\eta_{th} = \frac{1/2 \left(V_{10}^2 - V^2 \right)}{f H_f}. \tag{4.10}$$

The engine thermal efficiency is similar to the definition of the combustion chamber efficiency, but it includes all the engine components. Therefore energy not released in the combustion chamber because of inefficiencies of the mixing and combustion processes appears as the engine thermal efficiency.

4.5 Combined Cycles

One way to avoid expendable staging and make use of more efficient engine cycles during part of the ascent to orbit is by using two or more separate propulsion systems that operate independently on the vehicle. These are referred to as *combination propulsion systems* (CPSs). An example of this type of propulsion system is the rocket–ramjet, which uses a rocket booster to achieve the initial acceleration to a speed capable of sustaining ramjet operation (Billig, 1996a). At that speed, the engine switches to ramjet operation for the remainder of the flight. Although the use of a CPS simplifies propulsion system integration issues, it requires carrying at least one propulsion system that is not actively participating in propelling the vehicle at all times and thus leads to inefficient use of weight and volume and increases the heating load.

Another way to use high-efficiency air-breathing cycles during ascent in a reusable system is through the use of CCP systems. CCP systems can be broadly divided into two categories: air-breathing combined cycles that could include turbojet or turbofan cycles, and combined-cycle systems that include a rocket subsystem. Examples of air-breathing CCP systems are the dual-mode combustion ramjet, which operates in both ramjet and scramjet modes (Curran et al., 1996), and the turbine-based combined-cycle engine, which uses a turbine-based cycle for low-speed flight along with ramjet and scramjet modes (Georgiadis et al., 1998). Air-breathing combined-cycle engines are intended primarily for missions involving high-speed cruise, in the atmosphere, but cannot support transatmospheric flight when the air density becomes too low to sustain the cycle. A rocket-based cycle is then needed. The following subsections discuss some of the relevant issues for the combined cycles by use of a ramjet–scramjet architecture in combination with gas-turbine or rocket engines.

4.5.1 The Turbine-Based Combined Cycle – TBCC

TBCCs are particularly attractive for the unsurpassed specific impulse at take-off. In that regard, TBCCs are of particular interest for TSTO concepts in which the first stage spends most of its mission at a relatively low supersonic speed. Recent advances concentrate on the development of turbine engine technologies that could operate efficiently up to Mach 4 (Bartolotta et al., 2003).

A simple combined cycle is a turbojet (or turbofan)–ramjet in which a secondary flow bypasses the core turbojet and participates to thrust in an afterburner. As the Mach number increases typically beyond $M = 3$, the afterburner transitions to operation as a ramjet and the turbojet maximum cycle temperature is reduced to maintain the load on the rotating machinery while maintaining the airflow path open to contribute to thrust generation in the afterburner. As is evident, the main issues are the matching of the flows through the core and through bypass and to avoid reversed flow. Additional operational difficulties are introduced by the broad bypass-ratio range during acceleration and deceleration and thermal management of the moving parts at high-enthalpy flight.

An interesting combined cycle that includes turbojet–rocket interaction is described by the KLIN^{TM} cycle (Balempin et al., 2002) shown schematically in Fig. 4.5. The rocket fuel, hydrogen at high pressure, is used to provide deep cooling of the air in the turbojet intake. In the diagram in Fig. 4.5, both the rocket and the turbojet use hydrogen as the fuel but the two fuel circuits could be separated to use nonsimilar fuels, for example, liquid hydrocarbons, for the turbojet. Although the rocket and the turbojet use different flow paths, there is a close interaction between the cycles because the rocket's fuel is used to cool the turbojet's incoming air to increase the density and reduce the temperature, thereby increasing the compression in the turbojet and extending its operation to higher Mach numbers. As the velocity changes, both the turbojet

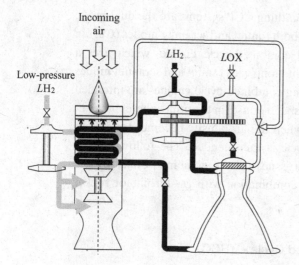

Figure 4.5. The KLIN™ turbojet–rocket configuration (Balempin et al., 2000) incorporates a deep-cooled turbojet cycle along with a liquid-fuel rocket. Both the turbojet and the rocket operate with hydrogen as fuel.

and the rocket are throttled to adapt to both the low- and the high-speed regimes. This system is estimated to have up to Mach 6.5 operational capability.

Although it is not strictly a combined cycle, the liquid–air collection engine (LACE) is a combination of rocket cycles with air collection in flight through an arrangement as shown in Fig. 4.6. The concept involves air collection during initial stages of rocket operation through an atmosphere, chilled by liquid hydrogen, and the condensed liquid oxygen is injected in the main engine, where it burns with the liquid hydrogen. The propulsion system's overall weight is thereby reduced. The concept was included in the design of the British HOTOL program in the mid-1980s (Hallion, 1995).

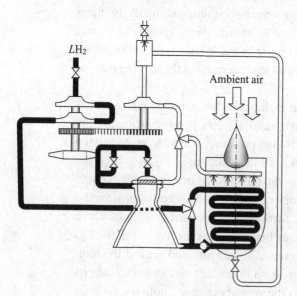

Figure 4.6. LACE concept using mixed air and an oxygen oxidizer (Balempin et al., 2000).

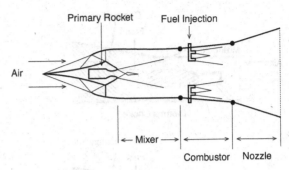

Figure 4.7. Schematic diagram of a RBCC with a rocket acting as an ejector to augment the airflow into the ramjet–scramjet.

4.5.2 The Rocket-Based Combined Cycle – RBCC

Among the many types and variations of CCP systems (Daines and Segal, 1998), one class of rocket-based CCP systems shows particular promise for Earth-to-orbit missions. These are engines that operate in rocket–ejector mode and also have the capability of operating in ramjet, scramjet, and rocket-only modes; they are typically referred to as rocket-based combined-cycle (RBCC) engines. One variant is the ejector scramjet engine shown schematically in Fig. 4.7. This concept was identified as one of the most promising propulsion systems for both SSTO and TSTO vehicles (Escher and Flornes, 1966).

The ability to utilize the rocket as an ejector increases the engine mass flow and therefore thrust. Afterburning in rocket–ejector mode, using the ramjet–scramjet fuel injectors, further increases the thrust and specific impulse compared with the rocket operating alone. As the ratio of the bypass air to the rocket exhaust mass flow increases with increasing flight speed, the specific impulse continues to increase as the cycle more closely resembles ramjet operation. In ramjet and scramjet modes, the rocket could be advantageously used as a fuel injector and mixing enhancer. In the rocket-only mode, the use of the engine duct as a highly expanded nozzle at high altitudes increases the specific impulse of that mode of operation.

A further advantage of RBCC systems is the reduction in the amount of onboard oxidizer required. This decreases the size and therefore the weight of the tank and vehicle. Vehicle propellant mass fractions for RBCC-powered vehicles are projected to be around 70%, compared with 90% for all-rocket vehicles (Escher et al., 1995). In the rocket–ejector mode, RBCC systems can provide vehicle thrust-to-weight ratios greater than one and are therefore capable of vertical takeoff and landing. Finally, the cryogenic fuel can be used in air-breathing modes as a heat sink to increase the density of the inlet airflow, thus increasing the work output. In terms of the thermodynamic cycle, this is equivalent to a more efficient process between Stations 0 and 2.1 shown in Fig. 4.3.

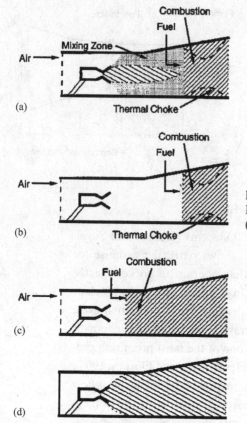

Figure 4.8. Operation of an ejector scramjet RBCC: (a) rocket–ejector, (b) ramjet, (c) scramjet, (d) rocket only.

4.5.2.1 RBCC Systems' Mode of Operation

As one of the most promising RBCC configurations the ejector scramjet shown in Fig. 4.7 is the basis for an entire class of RBCC engines. It consists of a rocket subsystem incorporated in an air-breathing engine and an inlet, mixer, combustion chamber, and nozzle. Fuel-injection sites can be located at several locations along the duct to optimize the fuel-injection selection according to the requirements of the flight regime and engine operation. The ejector scramjet operates in the four modes described in Fig. 4.8: rocket–ejector, ramjet, scramjet, and rocket-only mode.

The rocket–ejector mode shown in Fig. 4.8(a) is an ejector cycle with the rocket acting as the primary or drive jet. The thrust of the rocket is augmented through a jet-pumping process that transfers momentum from the high-velocity rocket exhaust to the inducted air. The ejector process results in an increased total mass flow with a lower exit velocity and yields a higher specific impulse in comparison with the rocket-only operation. The rocket–ejector mode is used from takeoff through low supersonic flight speeds. The specific impulse is typically augmented by 10%–20% under static conditions, and the

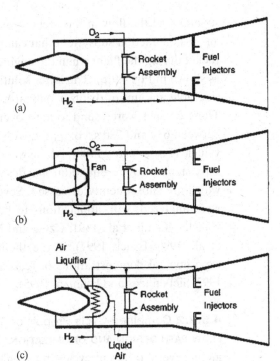

Figure 4.9. Schematic diagram of subsystems that could be added to an ejector scramjet engine: (a) basic ejector scramjet, (b) ejector scramjet with turbofan, and (c) ejector scramjet with air-liquefaction system.

augmentation increases to levels up to 250% at Mach numbers between 2 and 3. Much of the thrust augmentation is accomplished in the rocket–ejector mode by afterburning fuel with the inducted air in the duct downstream of the rocket (Dykstra et al., 1997). As the flight Mach number approaches 3, the engine transitions to ramjet mode [Fig. 4.8(b)], which provides a higher specific impulse in the mid- to high-supersonic flight-speed range. Oxidizer is supplied by the ram air from the inlet, and combustion takes place at subsonic conditions. Around $M = 6$, the operation of the engine turns to the scramjet mode [Fig. 4.8(c)], and the flow remains supersonic throughout the entire engine. The engine combustion cross section must remain constant or diverge in this mode to avoid the onset of thermal choking in the scramjet. The rocket is either turned off or used as a fuel injector in both ramjet and scramjet modes. Around $M = 15$ the air density can no longer sustain an efficient air-breathing cycle, and the engine is switched to the rocket-only operation, as shown in Fig. 4.8(d). The air inlets close and the rocket restarts, providing thrust to insert the spacecraft into orbit.

Several extensive studies (Escher and Flornes, 1966; Foster et al., 1989) evaluated a number of engine configurations for applications including SSTO and TSTO vehicle concepts. Among these configurations, the most promising that emerged consisted of the basic ejector scramjet shown in Fig. 4.8(a) with one or more additional subsystems. For example, the basic ejector scramjet cycle engine shown in Fig. 4.9(a) can be complemented by a turbofan to supercharge the flow in the rocket–ejector mode, as shown in Fig. 4.9(b), or an air-liquefaction subsystem that can produce the necessary oxidizer for the rocket during the flight when the engine operates in the rocket–ejector mode, as shown in Fig. 4.9(c). The latter solution eliminates the need to carry a considerable amount of oxidizer onboard, resulting in a reduced vehicle weight. These engines were found to have overall mission-effective specific impulses between 630 and 780 s, compared with 370 s for a dual-fuel, all-rocket SSTO vehicle (Foster et al., 1989). Along with added capabilities of the engine, these subsystems also present additional design challenges for successful operation, which are subsequently discussed. Several other vehicles with CCP systems were analyzed, with applications to both multiple-staged and single-staged vehicles (Ganji et al., 1991; Czysz and Murthy, 1995; Billig, 1996b; Sosounov et al., 1996; Esher, 1997). These theoretical studies were accompanied by experimental demonstrations of feasibility and mode transitions (Leingang, 1992; Siebenhaar and Bulman, 1995).

4.5.2.2 Combined-Cycle Propulsion Technical Issues

FLOW-PATH DESIGN AND OPTIMIZATION. The advantage of being able to operate in several different cycles in a single engine carries with it the additional requirement of designing a flow path that will provide an acceptable performance in each operational mode. The inlet will need to operate with a very low contraction in the rocket–ejector mode to capture as much air as possible (Billig, 1993). However, in the scramjet mode it will need to have a large enough contraction to provide sufficient compression of the incoming air before combustion. The optimum exit flow path in the rocket–ejector and the ramjet modes includes a converging–diverging section, whereas the scramjet mode requires straight and diverging sections only. Variable geometry, although an obvious solution, would add significant weight and complexity to the engine (Rohde, 1992). Fixed-geometry flow paths are possible through the use of thermal compression and thermal choking to provide an effect analogous to area change in the flow path. Tailoring of the fuel-injection location and amount is used to alter the flow instead of variable geometry and requires careful design of the fuel-injection system. Fixed-geometry inlets using thermal compression were proposed by Ferri in 1973, and significant performance enhancement was shown to be possible at low hypersonic speeds (Billig et al., 1968). Newer concepts based on MHD energy extraction in the inlet and redistribution in the nozzle (Burakhanov et al., 2001) point as well to the possibility of adjusting the inlet flow to the flight conditions without geometrical changes. However, progress on these concepts has been limited by the difficulty involved in tailoring the flow, fuel injection, heat release, and vehicle integration (Curran et al., 1996). Furthermore, for a fixed combustor–nozzle geometry, the flow path would also need to be optimized to allow controllable thermal choking in the rocket–ejector and the ramjet modes and to avoid thermal choking in the scramjet mode.

FUEL SELECTION AND DENSIFICATION. The issue of fuel system selection is an important integrating factor in the development of high-speed propulsion systems, including combined-cycle approaches. It encompasses issues of fuel management, stability, and energy density, along with the need for fast breakup and chemical decomposition of the injected fuel. Often these requirements are in contradiction, because high-energy–density fuels require high activation energies to initiate exothermic reactions (Segal and Shyy, 1996). For SSTO vehicles, hydrogen provides an overall specific impulse better than hydrocarbon-based fuels because of the higher energy–density and provides a source for active cooling of the airframe. In addition, the fast chemical kinetics of hydrogen contribute to reducing the combustion time in the scramjet mode operation. Advances such as gelled hydrogen (Palaszewski et al., 1997) or slush hydrogen (Escher, 1992) provide methods to increase the density of hydrogen. Slush hydrogen yields a 15% increase in density compared with that of liquid hydrogen, and it also provides 20% greater thermal sink. This is important, particularly in the LACE concept in which hydrogen "recycling," i.e., returning some hydrogen to the slush hydrogen tank for recooling, can increase the engine performance. For TSTO vehicles, the use of hydrocarbon-based fuels, including some newly formulated synthetic fuels with a high-energy content (Segal et al., 1995), is a possibility. A number of synthetic fuels were developed recently (Marchand et al., 2002) that have the potential of an increased gravimetric energy output, hence improving the vehicle mass properties. This category includes energetic fuels, including strained-bond molecules and hydrocarbons with large molecular formulations or those including azido groups, as well as solutions of more traditional formulations with energetic additives. Aspects of the combustion characteristics of several such energetic fuels were reviewed by Segal and Shyy (1996), Marchand et al. (2002), and Yang and Zarko (1995).

4.5.2.3 Mode-Specific RBCC Technical Issues

MIXING ENHANCEMENT IN ROCKET–EJECTOR MODE. When a single circular cross-sectional centerline-mounted rocket is used for rocket–ejector configurations, mixing lengths are large. Experimentally derived correlations for this configuration indicate that a duct length-to-diameter (L/D) ratio of 8 to 10 is required for complete mixing (Dykstra et al., 1997). Decreasing the duct length is important for reducing the engine weight; however, it cannot be accepted at the expense of incomplete mixing. Increasing the interfacial shear area between the primary and secondary flows increases mixing and reduces the required length because mixing results primarily from the turbulent and viscous shear forces in steady-flow ejectors. Therefore using a larger number of smaller primary rockets has proven effective in reducing mixing length (Siebenhaar and Bulman, 1995; Gregory and Han, 2003). An annular bell rocket has been suggested with a toroidal combustion chamber and an annular nozzle that increases the shear area (Escher and Schnurstein, 1993). It has been shown (Daines and Merkle, 1995) that an ejector utilizing an annular bell rocket mixes about four times as fast, lengthwise, as an ejector with an on-axis primary jet, and it has been estimated that a dual concentric annular bell would have an L/D of ~1 for complete mixing.

Mixing can also be enhanced in an rocket–injector mode cycle by inducing large-scale motion between the primary and secondary streams, which effectively increases the shear area. Forced mixer lobes (Presz et al., 1988) and primary jets with noncircular cross sections (Ho and Gutmark, 1987; Kim et al., 1998) induce large-scale fluid motion through vortex formation. For highly elliptic-shaped jets, the entrainment of a secondary fluid on the minor axis is increased by as much as a factor of eight compared with a circular jet (Liou et al., 1993), whereas the mixing rate on the major axis remains similar to that of a circular jet.

Turbulent mixing, which occurs in steady-flow ejectors, increases the stagnation pressure losses in the flow and results in a lower performance compared with that of a theoretical ideal mixing. In contrast, dynamic ejectors rely primarily on unsteady-pressure waves to accelerate the secondary flow and accomplish the momentum transfer and can therefore perform better than steady-flow ejectors. For example, an intermittent jet ejector (Lockwood and Patterson, 1962), in which the primary jet is pulsed, resulted in 90% thrust augmentation compared with 30% augmentation for the corresponding steady-flow ejector. Resonant acoustic modes excited naturally by the primary jet in some ejectors were correlated with increased mass entrainment (Bowman et al., 1990). Other unsteady ejector modes were suggested to improve mixing, including (i) rotary jets, in which the primary jets emanating from a freely rotating cylindrical or annular rotor drive the secondary air through the engine (Amin and Garris, 1995), and (ii) switching the rocket exhaust flow from side to side in a planar rocket duct to increase acceleration of the slower secondary air (Bulman, 1993); computational results indicated an increase of over 30% in specific impulse and a mass entrainment increase of over 10% at a switching frequency of 500 Hz compared with a steady-flow ejector in the latter configuration (Daines and Bulman, 1996). Although dynamic ejectors may prove useful in combined-cycle engines, practical technical issues such as increased weight, induced vibrations, and achieving jet switching must first be resolved.

SIMULTANEOUS MIXING AND COMBUSTION VERSUS DIFFUSION AND AFTERBURN-ING. Related to the issue of enhanced mixing is the question of whether to use diffusion and afterburning (DAB) or simultaneous mixing and combustion (SMC) for the afterburning in the rocket–ejector mode. In the SMC approach, fuel-rich rocket exhaust is used to drive the mixer flow and combustion is allowed to occur simultaneously with mixing and expansion. The resulting subsonic flow stream is then passed through a converging–diverging nozzle and expanded to supersonic velocities. An alternative approach is to mix a stoichiometric supersonic rocket drive jet with the subsonic inlet airstream and expand the combined subsonic flow stream to increase the static pressure. At the peak pressure point, additional fuel is introduced and burned, and the entire flow is expanded through a converging–diverging nozzle. This approach is referred to as DAB. The SMC cycle exhibits a consistently lower engine specific impulse at low Mach numbers relative to DAB cycles, as one would expect from the basic thermodynamic consideration governing heat engine cycle efficiency. This difference is significant at sea-level static conditions but diminishes progressively with increasing Mach number. One experimental study (Stroup and Pontzer, 1968) showed that combustion efficiency of the afterburner in the rocket–ejector mode decreased from over 90% with DAB to about 40% with SMC by decreasing the length available for mixing before fuel injection. However, a SMC engine with a fuel-rich rocket exhaust has the advantage that a separate downstream fuel-injection capability is unnecessary, thereby reducing the engine weight and complexity. Billig (1993) suggested that a shorter engine duct more than offsets the lower efficiency by a compensatory decrease in engine weight. Furthermore, one suggested method (Siebenhaar and Bulman, 1995) to minimize losses is to introduce a fuel-rich flow that is shielded by the rocket exhaust from immediately mixing and allowed to react with the secondary air. This eliminates the need for downstream fuel injection, while allowing improved mixing before afterburning occurs.

ROCKET-ONLY MODE CYCLE EFFICIENCY. RBCC systems can make use of the air-breathing duct to act as a high-expansion nozzle when the ambient pressure is low to increase the overall performance. This rocket-only mode of operation has to be considered during the flow-path optimization because a well-designed ramjet or scramjet flow path does not necessarily result in high efficiency in the rocket-only mode. The study by Steffen et al. (1998) evaluated the effect of various parameters, including the engine duct area at the rocket exit plane, rocket nozzle exit area, wall angle, and base bleed on the cycle efficiency of an RBCC engine in rocket-only-mode operation. Results showed that a large engine duct area at the rocket exit plane and a long engine duct resulted in a decreased specific impulse whereas a large rocket nozzle exit area and engine duct exit area increased the specific impulse. In addition, for a divergent nozzle, base bleed reduced the specific impulse. Depending on the geometry, cycle efficiencies ranged from about 78% to 95% of ideal rocket performance, which was computed assuming a well-designed nozzle with the same overall expansion ratio.

ENHANCEMENTS TO THE BASIC EJECTOR SCRAMJET CONFIGURATION. System studies indicated several subsystems could be added to the basic ejector scramjet to increase the specific impulse. One subsystem that improves the specific impulse in the rocket–ejector mode is a turbofan included in the flow path before the rocket, as shown in Fig. 4.9(b). A turbofan adds the capability of powered loiter with a substantially increased specific impulse, as high as 23 000 by some estimates (Escher, 1997). However, these advantages come at the expense of increased installed weight and complexity. A major issue with this option is the removal of the rotating machinery from the flow path and stowage during elevated-Mach-number flight to protect it from the extreme temperature conditions that would be experienced. Several methods have been suggested, including swinging or rotating the fan out of the flow path (Escher, 1997).

A method suggested for increasing the specific impulse in a rocket–ejector mode at the expense of extra weight is to include a LACE subsystem, shown schematically in Fig. 4.9(c), which implements in situ air liquefaction to provide the oxidizer for the rocket (Escher, 1992; National Research Council, 1998). LACE systems have the advantage of further reducing the volume of the stored oxidizer and therefore reducing the oxidizer tank size and weight. This type of engine collects and liquefies a portion of the incoming air in a heat exchanger that utilizes liquid hydrogen fuel in the condenser. Use of this subsystem would require a very compact, lightweight heat exchanger and a method for alleviating fouling and icing in the heat exchanger. In addition, more hydrogen is required for liquefying the air than is necessary for stoichiometric engine operation, which would result in fuel-rich operation that thereby decreases the specific impulse unless a thermal sink, such as slush hydrogen, is provided to recycle the excess fuel.

REFERENCES

Amin, S. M. and Garris, C. A. J. (1995). "An experimental investigation of a non-steady flow thrust augmenter," AIAA Paper 95–2802.

Anderson, G. Y., McClinton, C. R., and Weidner, J. P. (2000). "Scramjet performance," in *Scramjet Propulsion* (E. T. Curran and S. N. B. Murthy, eds.), Vol. 189 of Progress in Astronautics and Aeronautics, AIAA.

Balempin, V. V., Maita, M., and Murthy, S. N. B. (2000). "Third way of development of single-stage-to-orbit propulsion," *J. Propul. Power* **16**, 99–104.

Balempin, V. V., Liston, G. L., and Moszée, R. H. (2002). "Combined cycle engines with inlet conditioning," AIAA Paper 2002–5148.

Bartolotta, P. A., McNellis, N. B., and Shafer, D. G. (2003). "High speed turbines: Development of a turbine accelerator (RTA) for space access," AIAA Paper 2003–6943.

Billig, F. S. (1993). "The integration of the rocket with the ram-scramjet as a viable transatmospheric accelerator," International Society for Air Breathing Engines, Paper 93–7017.

Billig, F. S. (1996a). "Tactical missile design concepts," in *Tactical Missile Propulsion* (G. E. Jensen and D. W. Netzer, eds.), Vol. 170 of Progress in Astronautics and Aeronautics, AIAA.

Billig, F. S. (1996b). "Low-speed operation of an integrated rocket-ram-scramjet," in *Developments in High-Speed Vehicle Propulsion Systems* (S. N. B. Murthy and E. T. Curran, eds.), Vol. 165 of Progress in Astronautics and Aeronautics, AIAA, pp. 51–104.

Billig, F. S., Orth, R. C., and Lasky, M. (1968). "Effects of thermal compression on the performance estimates of hypersonic ramjets," *J. Spacecr. Rockets* **5**, 1076–1081.

Bowcutt, K. G. (2001). "Multidisciplinary optimization of airbreathing hypersonic vehicles," *J. Propul. Power* **17**, 1184–1190.

Bowman, H., Gutmark, E., Schadow, K., Wilson, K., and Smith, R. (1990). "Supersonic rectangular isothermal shrouded jets," AIAA Paper 90–2028.

Bryson, A. E., Desai, M. N., and Hoffman, W. C. (1969). "Energy-state approximation in performance optimization of supersonic aircraft," *J. Aircr.* **6**, 481.

Bullock, R. O. (1989). "Design and development of aircraft propulsion systems," in *Aircraft Propulsion Systems Technology and Design* (G. C. Oates, ed.), AIAA Educational Series, pp. 3–101.

Bulman, M. J. (1993). "Ejector ramjet," U.S. patent 5,205,119.

Burakhanov, B. M., Likhachev, A. P., Medin, S. A., Novikov, V. A., Okunev, V. I., Rickman, V. Yu., and Zeigarnik, V. A. (2001). "Advancement of scramjet magnetohydrodynamic concept," *J. Propul. Power* **17**, 1247–1252.

Czysz, P. and Murthy, S. N. B. (1995). "Energy management and vehicle synthesis," in *Developments in High-Speed-Vehicle Propulsion Systems* (S. N. B. Murthy and E. T. Curran, eds.), Vol. 165, of Progress in Astronautics and Aeronautics, AAIA.

Curran, E. T., Heiser, W. H., and Pratt, D. T. (1996). "Fluid phenomena in scramjet combustion systems," *Annu. Rev. Fluid Mech.* **28**, 323–360.

Daines, R. and Bulman, M. (1996). "Computational analyses of dynamic rocket ejector flowfields," AIAA Paper 96–2686.

Daines, R. and Merkle, C. (1995). "Computational analysis of mixing and jet pumping in rocket ejector engines," AIAA Paper 95–2477.

Daines, R. and Segal, C. (1998). "Combined rocket and airbreathing propulsion systems for space-launch applications," *J. Propul. Power* **14**, 605–612.

Dykstra, F., Caporicci, M., and Immich, H. (1997). "Experimental investigation of the thrust enhancement potential of ejector rockets," AIAA Paper 97–2756.

Escher, W. J. D. (1992). "Cryogenic hydrogen-induced air-liquefaction technologies for combined-cycle propulsion applications," NASA CP-10090, pp. 1–19.

Escher, W. J. D., ed. (1997). *The Synerjet Engine*, Progress in Technology, PT-54, Society of Automotive Engineers.

Escher, W. J. D. and Flornes, B. (1966). "A study of composite propulsion systems for advanced launch vehicle applications," Marquardt Corp., NAS7–377 Final Rept., Van Nuys, CA.

Escher, W. J. D., Hyde, E., and Anderson, D. (1995). "A user's primer for comparative assessments of all-rocket and rocket-based combined-cycle propulsion systems for advanced earth-to-orbit space transportation," AIAA Paper 95–2474.

Escher, W. and Schnurstein, R. (1993). "A retrospective on early cryogenic primary rocket subsystem designs as integrated into rocket-based combined-cycle (RBCC) engines," AIAA Paper 93–1944.

Ferri, A. (1973). "Mixing controlled supersonic combustion," *Annu. Rev. Fluid Mech.* **5**, 301–338.

Foster, R., Escher, W., and Robinson, J. (1989). "Studies of an extensively axisymmetric rocket based combined cycle (RBCC) engine powered SSTO vehicle," AIAA Paper 89–2294.

Ganji, A. R., Khadem, M., and Khandani, S. M. (1991). "Turbo/air-augmented rocket: A combined cycle propulsion system," *J. Propul. Power* **7**, 650–653.

Georgiadis, N. J., Walker, J. F., and Trefny, C. J. (1998). "Parametric studies of the ejector process within a turbine-based combined-cycle propulsion system," AIAA Paper 98–0936.

Gregory, D. C. and Han, S. (2003). "Effects of multiple primary flows on ejector performance in an ejector-ram rocket engine," AIAA Paper 2003–373.

Guoskov, O. V., Kopchenov, V. I., Lomkov, K. E., Vinogradov, V. A., and Waltrup, P. J. (2001). "Numerical researches of gaseous fuel pre-injection in hypersonic 3-D inlet," *J. Propul. Power* **17**, 1162–1169.

Hallion, R. P. (1995). *The Hypersonic Revolution, Volume II: From Max Valier to Project Prime*, Aeronautical System Center, Air Force Material Command, Wright Patterson Air Force Base, ASC-TR-95–5010.

Hargraves, C. R. and Paris, S. W. (1987), "Direct trajectory optimization using nonlinear programming and collocation," *J. Guidance* **10**, 338–342.

Heiser, W. H. and Pratt, D. T. (with Daley, D. H. and Mehta, U. B.). (1994). *Hypersonic Airbreathing Propulsion*, AIAA Educational Series.

Ho, C.-H. and Gutmark, E. (1987). "Vortex induction and mass entrainment in a small-aspect-ratio elliptic jet," *J. Fluid Mech.* **179**, 383–405.

Kim, J.-H., Samimy, M., and Erskine, W. R. (1998). "Mixing enhancement with minimal thrust loss in a high speed rectangular jet," AIAA Paper 98–0696.

Leingang, J. (1992). "Advanced ramjet concepts program," NASA CP-10090, pp. 1–9.

Liou, T., Chen, L., and Wu, Y. (1993). "Effects of momentum ratio on turbulent non-reacting and reacting flows in a ducted rocket combustor," *Int. J. Heat Mass Transfer* **36**, 2589–2599.

Lockwood, R. and Patterson, W. (1962). "Energy transfer from an intermittent jet to a secondary fluid in an ejector-type thrust augmenter," Hiller Aircraft Co., Rept. ARD-305, Newark, CA.

Marchand, A. P., Kruger, H. G., Power, T. D., and Segal, C. (2002). "Synthesis and properties of polycyclic cage hydrocarbons (high energy density fuels) and of polycyclic cage organonitro compounds (insensitive high-energy explosives)," *Kem. Ind.* **51**, 51–67.

Mattingly, J. D., Heiser, W. H., and Pratt, D. T. (2002). *Aircraft Engine Design*, 2nd ed., AIAA Educational Series.

Mehta, U. B. and Bowles, J. V. (2001). "Two-stage-to-orbit space plane concept with growth potential," *J. Propul. Power* **17**, 1149–1161.

Mitani, T., Hiraiwa, T., Tarukawa, Y., and Masuya, G. (2002). "Drag and total pressure distribution in scramjet engines at Mach 8 flight," *J. Propul. Power* **18**, 953–960.

National Research Council. (1998). "Maintaining US leadership in aeronautics: Breakthrough technologies to meet future air and space transportation and goals," National Academy Press.

Oran, E. S. and Boris, J. P. (2001). *Numerical Simulation of Reactive Flow*, 2nd ed., Cambridge University Press.

Ortwerth, P. J. (2000). "Scramjet flowpath integration," in *Scramjet Propulsion* (E. T. Curran and S. N. B. Murthy, eds.), Vol. 189 of Progress in Astronautics and Aeronautics, AIAA.

Palaszewski, B., Ianovski, L. S., and Carrick, P. (1997). "Propellant technologies: A persuasive wave of future propulsion benefits," in *Proceedings of the Third International Symposium on Space Propulsion* (V. Yang, G. Liu, W. Anderson, and M. Habiballah, eds.), Chinese Society of Astronautics, pp. 1–13.

Presz, W., Morin, B., and Gousy, R. (1988). "Forced mixer lobes in ejector designs," *J. Propul. Power*, **4**, 350–354.

Riggins, D. W. (1997). "Thrust losses in hypersonic engines, part 2: Applications," *J. Propul. Power* **13**, 288–295.

Rohde, J. (1992). "Airbreathing combined cycle engine systems," NASA CP-10090, pp. 1–9.

Schmidt, D. K. (1997). "Optimum mission performance and multivariable flight guidance for airbreathing launch vehicles," *J. Guid. Control Dyn.* **20**, 1157–1164.

Schmidt, D. K. and Hermann, J. A. (1998). "Use of energy-state analysis on a generic air-breathing hypersonic vehicle," *J. Guid. Control Dyn.* **21**, 71–76.

Segal, C., Friedauer, M. J., Udaykumar, H. S., and Shyy, W. (1995). "Combustion of high-energy fuels in high speed flows," in *Proceedings of the Eighth ONR Propulsion Meeting* (G. D. Roy and F. A. Williams, eds.), University of California, San Diego, pp. 192–199.

Segal, C. and Shyy, W. (1996). "Energetic fuels for combustion applications," *J. Energy Resources Technol.* **118**, 180–186.

Semenov, V. L., Prokhorov, A. N., Strokin, M. V., Relin, V. L., and Alexandrov, V. Yu. (2002). "Fire tests of experimental scramjet in free stream in continuously working test facility," available at http://hypersonic2002.aaaf.asso.fr/papers/Semenovpaper.pdf.

Siebenhaar, A. and Bulman, M. (1995). "The strutjet engine: The overlooked option for space launch," AIAA Paper 95–3124.

Sosounov, V. S., Tskhovrebov, M. M., Solonin, V. I., Kadjardousov, P. A., and Palkin, V. A. (1996). "Turboramjets: Theoretical and experimental research at Central Institute of Aviation Motors," in *Developments in High-Speed Vehicle Propulsion Systems* (S. N. B. Murthy and E. T. Curran, eds.), Vol. 165 of Progress in Astronautics and Aeronautics, AIAA, pp. 205–258.

Steffen, C. J., Jr., Smith, T. D., Yungster, S., and Keller, D. J. (1998). "Rocket based combined cycle nozzle analysis using NPARC," AIAA Paper 98–0954.

Stroup, K. and Pontzer, R. (1968). "Advanced ramjet concepts, Volume I: Ejector ramjet systems demonstration," Marquardt Corp., Air Force Aero Propulsion Lab., TR-67–118.

Trefny, C. J. (1999). "An air-breathing launch vehicle concept for single-stage-to-orbit," NASA TM-209089.

Vinogradov, V., Segal, C., Owens, M., and Mullargili, S. (2001). "Effect of kerosene preinjection on combustion flameholding in a Mach 1.6 airflow," *J. Propul. Power* **17**, 605–611.

Warnatz, J., Maas, U., and Dibble, R. W. (1996). *Combustion – Physical and Chemical Fundamentals, Modeling and Simulation, Experiments, Pollutant Formation*, Springer-Verlag.

Yang, V. and Zarko, V. E., eds. (1995). "Solid propellant rocket motor interior ballistics and combustion of energetic materials," special issue of *J. Propul. Power* **11**(4).

Younghans, J. L. (1989). "Inlets and inlet/engine integration," in *Aircraft Propulsion Systems Technology and Design* (G. C. Oates, ed.), AIAA Educational Series, pp. 291–298.

5 Inlets and Nozzles

5.1 Inlets

5.1.1 Introduction

Air intakes for any air-breathing engine-equipped vehicles must

> capture the exact amount of air required by the engine, accomplish the deceleration to the required engine entrance air speed with minimum total pressure loss, deliver the air with tolerable flow distortion and contribute the least possible drag to the system (Mahoney, 1990).

These general requirements for all air-breathing engine inlets would place particular emphasis on some of the stated functions or others, depending on the specific characteristics of the propulsion system used and the vehicle mission. Some of these requirements are of general applicability; minimum pressure losses and least possible drag induction fall into this category. Other inlet characteristics have more or less significant influence, depending on the particular engine used. For example, dynamic distortions induced by an inlet can create serious difficulties for a gas-turbine-engine compressor because they reduce the stall margin, thus limiting the operational range. The extent to which the dynamic distortions affect a scramjet engine operation, on the other hand, is not entirely clear because increased flow unsteadiness could accelerate mixing but may also have a negative effect on momentum losses. This is not the case for the steady-state flow nonuniformities that have been shown to cause significant effects on the scramjet flow field, as they do on other engines.

Design considerations derived from mission requirements lead to specific inlet characteristics. A duct offset, for example, which is required in many cases for a gas-turbine-engine-based high-speed vehicle to reduce the radar signature, often leads to flow separation and pressure losses; a scramjet engine is not likely to have this particular design requirement. Diverting and/or bleeding the boundary layer to reduce shock–boundary-layer interactions is often used for gas-turbine-based engines; it would be challenging to implement diverting boundary layers in a scramjet engine inlet because these boundary layers are thick after having developed along lengthy ramps and perhaps the entire vehicle forebody. Displacing the inlet from the fuselage to reduce boundary-layer ingestion, as is done in other high-speed vehicles, is not practical for the scramjet-equipped vehicle because of the unacceptable drag increase; besides, the hypersonic boundary layers are less likely to separate than the lower-speed boundary layers, and therefore their interaction with the shock system can be treated differently in practice.

Mechanical problems may pose significant challenges on the scramjet inlet design when compared with other inlet systems. Cooling may become necessary for flight at hypersonic speeds, and variable geometry – which is likely required to adapt to both low- and high-speed flight regimes – becomes more challenging under higher dynamic forces and elevated temperatures.

Finally, the range of operational conditions for the scramjet engine inlet is considerably larger than for the traditional supersonic vehicles. This remains true even when the complex flow interactions that exist in combined cycles are left out of this discussion; these interactions add yet another dimension to the inlet–propulsion system integration.

Adapting the inlet geometry to control the engine flow is increasingly required as the operational range of the vehicle increases. In current aircraft practice, fixed geometries satisfy the balance between complexity and weight on one side, with improved system performance in terms of compression efficiency, on the other, up to about Mach 2 flight conditions. Above this speed, the geometry must be allowed to change to adapt to flight conditions and engine power transients. The number of adjustable surfaces that maintain an optimal system of oblique shock wave and thus maximize the process efficiency multiplies as the speed increases and, along with it, the complexity and the inlet weight. For the scramjet that operates with only small changes around the design condition, such as the case of a hypersonic missile, fixed inlet geometries can provide adequate compression for the engine operation; however, a transatmospheric vehicle is likely to require some geometry adjustment to the broad range of flight conditions. This is particularly true as such vehicles are expected to include some combined cycle to operate efficiently from takeoff to orbit insertion. Given that the variable geometry is so specific to a particular system design and must be correlated with the nozzle design and operation, the following discussion is restricted to fixed inlet geometries only to make evident the physical flow processes relevant to all air intakes.

It is likely that the scramjet inlet (i) will compress the oncoming air using all of its surfaces, thereby resulting in a complex 3D shock-wave system in the duct preceding the combustion chamber entrance, (ii) will use variable geometry to accommodate the engine flow-rate requirements over the entire flight

Figure 5.1. HRE axisymmetric inlet model (Andrews et al., 1974). Designed for operation within the Mach 4–8 regime, this inlet has a triple-cone spike of 10°, 15.8°, and 22°, respectively. The spike translates to ensure "shock-on-lip" operation between Mach 6 and Mach 8. The outward spike translation is associated with increased internal geometric contraction. At Mach numbers below 6, the cone is fixed, resulting in increased spillage.

regime from supersonic to hypersonic, (iii) will provide compatibility with the combustion chamber pressure rise, (iv) will achieve a high degree of integration with the fuselage because of the long compression ramps it requires at hypersonic flight, and (v) will be arranged in a single segment or in several segments, depending on the size of the vehicle and the optimization of the propulsion system's frontal area. Additional design details must refer to aerodynamic heating treatment, shock–boundary-layer interactions, starting characteristics, and flow nonuniformity generation.

If the inlet generates a flow field that has a determining effect on the combustion chamber flow conditions, the combustor in turn may affect the inlet operation to a considerable degree. During "on-design" operation, the pressure rise that is due to heat release in the combustion chamber is not expected to lead to upstream interactions; however, during transients and, in particular, during acceleration, increased heat deposition in the combustion chamber may lead to upstream interactions and air mass flow reduction that would affect the inlet flow. A section inserted between the inlet and the combustion chamber, referred to as an "isolator," is needed to absorb these pressure differences. The isolator's operation is discussed in Chap. 6 in relation to the heat-release process in the combustion chamber.

Two examples of hypersonic inlets are given in Figs. 5.1 and 5.2. The first is an axisymmetric configuration used in the HRE model (Andrews et al., 1974; Andrews and Mackley, 1976), and the second is a rectangular configuration in which efficient compression within a short length with minimal flow turning is achieved through a complex shock-wave system on all four inlet sides as in the configurations studied by Holland and Perkins (1990) or Goldfeld et al. (2001). A variety of other configurations deriving from the mission requirements and vehicle architecture were studied, including 3D "alligator"-type, slit-side inlets that facilitate starting and improve operability in both subsonic and supersonic regimes (Hsia et al., 1991) and even Busemann-type designs optimized for a particular flight Mach number (Van Wie and Molder, 1992; Billig et al., 1999).

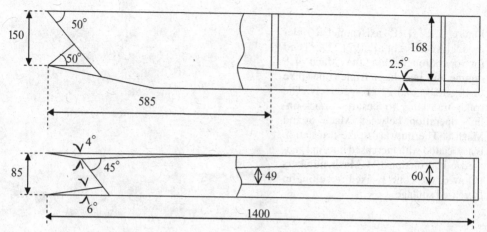

Figure 5.2. The 3D inlet (after Goldfeld et al., 2001).

The discussion that follows refers to energetic aspects of the inlet compression process, particular design considerations for scramjet operational conditions, and compatibility with the propulsion system and gives an overview of some advanced concepts suggested for flow manipulation in the inlet to improve the inlet–propulsion system performance.

5.1.2 Compression Process Efficiency and Energetic Balance

5.1.2.1 Pressure Recovery and Kinetic Energy Efficiency

As expected for high-speed flight, the scramjet engine's inlet will operate with both external and internal compression. Unlike other types of inlets, the intake system strictly serving as a scramjet – thus excluding from this discussion complex configurations needed for combined-cycle engines – does not include subsonic flow except for small regions confined to the boundary layer and possible separation regions. Slowing the flow to velocities that are reasonably low for flameholding and rapid heat release in the engine may result in a significant temperature rise; this, in turn, would reduce the amount of heat that can be generated efficiently during the combustion process. A balance must thus be achieved between the requirement to reduce the internal flow velocity to a range that is reasonable for stable combustion and the need to maintain high operational efficiency. It is therefore feasible that an optimal thermal compression ratio, ψ_{opt}, defined as the enthalpy ratio across the inlet (Ortwerth, 2000), could be formulated. This optimized compression ratio depends on the efficiency of the compression process, the departure from adiabatic conditions, the viscous effects, etc. The energetic aspects of the compression process, which are discussed later, can be related to the inlet thermal compression. Clearly the inlet flow is quite complex and only detailed 3D analyses can accurately describe it. However, simplified 1D analyses based on linear parameters,

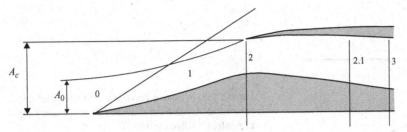

Figure 5.3. Inlet station nomenclature.

which may provide useful tools for preliminary analyses and optimization, deserve attention.

Figure 5.3 shows schematically selected inlet stations. It shows a case in which the captured mass flow is less than the maximum flow possible under ideal conditions of *"shock-on-lip"* design. Ingestion of the shock system generated by the vehicle forebody and the inlet ramps leads to a degradation of the inlet performance caused by undesirable and difficult-to-control viscous interactions. Therefore inlet designs allow, in general, a safety margin for shock ingestion and operate at the design point with A_0/A_c slightly less than unity. This is also the condition encountered during flight at speeds below the design point. The air contained in the flow tube between the free-stream air captured by the engine A_0 and the inlet entrance A_c is spilled around the cowl, resulting in an additive drag that must be included in the overall drag accounting.

Station 2 is the minimal inlet area, often referred to as a *throat* (Van Wie, 2000) by analogy with the external compression inlets. The diffuser follows between the throat and Station 2.1, which designates the entrance to the isolator, a component that is often treated separately from the inlet. Station 3 represents the combustion chamber entrance.

Figure 5.4 is an entropic diagram of the compression process in the inlet. The thermodynamic states are defined at the stations indicated in Fig. 5.3. A

Figure 5.4. Entropic diagram of the inlet compression process.

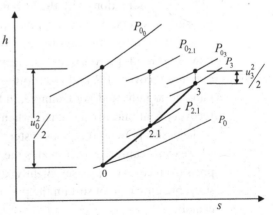

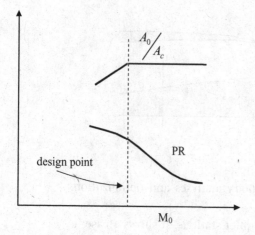

Figure 5.5. Pressure recovery and the inlet capture ratio dependence on the flight Mach number (after Mahoney, 1990).

continuous deceleration takes place in the inlet from the flight Mach number to Station 3, the entrance into the combustion chamber. Within the inlet–
isolator system, the air speed slows through the shock system and according to the duct area distribution. The nonisentropic effects result in a certain
amount of losses that continuously lower the stagnation pressure from one
station to another. The pressure recovery, defined as $PR = P_{t2.1}/P_{t0}$, is the primary parameter indicating the efficiency of the compression process. Stagnation pressure losses lead to reduced axial momentum and diminish the system performance. It is influenced by the flight Mach number and the engine
mass-flow requirement, effects of back-pressure and angle-of-attack changes.
Among all the factors that influence the pressure recovery, departures from
the design Mach number and engine flow rate contribute to the largest extent
to variations in inlet pressure recovery.

For a fixed-geometry inlet design, as the vehicle accelerates above the
design point, the shock system generated on the ramps leaves the cowl and
penetrates the inlet duct. This is a condition known as an "oversped" inlet.
The shock interactions with the flow field in the duct lead to an abrupt drop in
pressure recovery. The area capture ratio A_0/A_c remains unchanged. As the
vehicle decelerates, the capture ratio becomes less than unity, and, as the mass
flow is reduced, the pressure recovery generally increases. Qualitatively, the
pressure recovery PR and the capture ratio A_0/A_c vary with the flight Mach
number, as indicated, for example, in Fig. 5.5 (Mahoney, 1990). The pressure
recovery continuously decreases when the vehicle is flying at speeds in excess
of the design point as the shock system and the viscous losses become stronger,
whereas the capture ratio remains the same. Below the flight design point the
pressure recovery increases slightly, and instabilities may appear if the inlet
static pressure cannot sustain the pressure rise downstream in the combustion
chamber.

Along with the pressure recovery, additional inlet efficiency parameters can be defined to reflect the energy recovery (or loss) across the compression process. The adiabatic kinetic energy efficiency,

$$\eta_{KE_{ad}} = \frac{h_{t_0} - h(P_0, s_3)}{h_{t_0} - h_0},$$ (5.1)

is a measure of kinetic energy loss that is due to the entropy increase during the compression process and includes both the external and the internal compression. It assumes that the compression is done without heat exchange with the surroundings, and the static pressure is allowed to remain the same as in the free stream. When the nonadiabatic effects, which can be quite large for inlets operating at hypersonic conditions, are considered, the kinetic efficiency is reduced by a factor that accounts for the heat lost to the surroundings:

$$\eta_{KE} = \frac{h_{t_3} - h(P_0, s_3)}{h_{t_0} - h_0}.$$ (5.2)

Additional efficiency parameters can be formulated and are useful for describing static pressure rise and entropic and enthalpic airstream changes (Van Wie, 2000). If a fundamental thermodynamic relation $h = h(s, P)$ is known for the air captured by the scramjet inlet, relations between these various efficiency parameters can be derived. In particular, if the air is assumed thermally and calorically perfect, various efficiencies can be related through simple algebraic equations (Van Wie, 2000). As an example, the relation between the kinetic energy efficiency and the pressure recovery yields

$$\eta_{KE_{ad}} = 1 - \frac{2}{(\gamma - 1)M_0^2} \left[\left(\frac{1}{PR} \right)^{\gamma - \frac{1}{\gamma}} \right].$$ (5.3)

5.1.2.2 The Pressure Coefficient K_{WP}

If the viscous and shock losses are assumed to take place within a certain layer of the inlet airflow that has as a result a different average pressure than in the rest of the core flow, which is assumed to remain inviscid, a global pressure coefficient that incorporates the inlet losses can be defined as Ortwerth (2000) formulated it:

$$K_{WP} = \frac{\overline{P_W - P}}{P},$$ (5.4)

where the subscript W indicates that a layer that includes the pressure effects is differentiated from the inviscid pressure in the core. The overbar indicates that an average value is assumed for the entire inlet or for the inlet segment to which the analysis is applied. This coefficient is particularly useful for simplified 1D analyses because it can relate the inlet contraction ratio to the entropy rise in the inlet caused by various pressure-related losses; therefore the pressure coefficient K_{WP} can be used as a key coordinate reflecting the global energetic effects associated with the selected design. The connection results from a 1D momentum and energy analysis (Ortwerth, 2000) as follows:

$$V\frac{dV}{dx} = -\frac{1}{\rho}\left[\frac{dP}{dx} + (P_W - P)\frac{1}{A}\frac{dA}{dx} - \tau_W\frac{1}{A}\frac{d\sigma}{dx}\right], \tag{5.5}$$

$$\frac{dH}{dx} + V\frac{dV}{dx} = -\frac{q_W}{\dot{m}}\frac{d\sigma}{dx}, \tag{5.6}$$

where τ_W is the shear over an element of wall area σ, q_W is the heat lost to the same element of wall area, and $\dot{m}$ is the air mass flow. With the energy conservation written as

$$Tds = dH - \frac{dP}{\rho}, \tag{5.7}$$

the entropy dependence on the pressure coefficient becomes easily recognizable as

$$\frac{ds}{R} = -\left(\frac{P_W - P}{P}\right)\frac{dA}{A} + \left(\tau_W - \frac{q_W}{V}\right)\frac{1}{P}\frac{d\sigma}{A}. \tag{5.8}$$

Here, the losses that are due to the pressure deficiency resulting from compression become evident along with losses that are due to viscosity and heat transferred to the inlet structure. Thus Eq. (5.8) clearly emphasizes the pressure coefficient K_{WP} definition as the entropy increases because of area contraction. If the pressure coefficient can now be related to the inlet area contraction ratio and the flight Mach number, the thermodynamic state at the inlet exit plane can be directly determined. From extensive analyses with 2D inlets with single-edge ramps, Ortwerth (2000) suggested the following dependence of the pressure coefficient on the flight Mach number, contraction ratio, and compression-ramp angle:

$$K_{WP} = \frac{\omega(M_0)\alpha(M_0, \theta)}{\sqrt{\psi}}, \tag{5.9}$$

where ω and α are functions of the flight Mach number M_0 and the compression-ramp angle θ, and ψ is the thermal compression ratio. The Mach number functions ω and α are defined as

$$\omega(M_0) = \left(\frac{\gamma - 1}{\gamma}\frac{1}{M_0^2}\right), \tag{5.10}$$

the hypersonic similarity parameter α is defined as

$$\alpha(M_0, \theta) = \sqrt{M_0^2 - 1} \cdot \theta, \tag{5.11}$$

and the thermal compression ratio is as defined earlier, $\psi = h_3/h_1$.

It should be noted that this formulation of the thermal compression ratio refers to the inlet internal compression only, which can be related directly to the geometric contraction ratio, $CR = A_3/A_c$, as

$$\psi = h_3/h_1 = CR^{\gamma'-1} \tag{5.12}$$

The exponent γ' in Eq. (5.12) is a corrected specific heat ratio that accounts for the entropic changes associated with the pressure coefficient, K_{WP}, component,

$$\gamma' = \gamma + (\gamma - 1) K_{WP}. \tag{5.13}$$

Along with the kinetic energy efficiency η_{KE}, the pressure recovery PR, and the pressure coefficient K_{WP}, knowledge of the heat lost to the surroundings and wall shear stress leads to a complete set of information that allows the optimization of the inlet thermodynamic design.

5.1.2.3 Inlet Performance – Compression and Contraction Ratio Effects

Once the design point is selected in terms of flight conditions and a thermal contraction ratio is selected based on the desired combustion chamber conditions, the inlet design will further depend on the external and internal geometry, which can take a multitude of shapes according to the application, the structural restrictions, etc. Billig et al. (1999) suggested that a Busemann-type inlet, which offers a unique design solution, would still perform better at off-design conditions than an inlet that has not been optimally designed – in the sense of isentropic compression – at the design point. But even then, it was pointed out, the Busemann design can lead to unpractical solutions regarding length, weight, and viscous losses, and the design must be adapted – most often truncated to reduce weight and size – to the particular application.

It appears then that a parametric selection of the contraction ratio CR, along with the conceptual compression mechanism, i.e., 2D versus 3D or isentropic, number of compression surfaces, and an accurate account of the viscous losses and heat transferred to the walls, provides a design space that yields the compression ratio and the efficiency parameters needed to assess the design, pressure recovery, kinetic energy efficiency, and other factors.

An example of how the simple Busemann inlet design, shown in Fig. 5.6, responds to selected parameter changes is shown in Figs. 5.7 and 5.8; it was selected from the study by Billig et al. (1999). This inlet's flow field results from the isentropic compression initiated at the inlet capture and terminates with a conical shock having a half-cone angle θ_s. If an inviscid assumption is made, the Mach number remains constant throughout the rest of the inlet's constant-area duct because the shocks are canceled at Station 3. At an

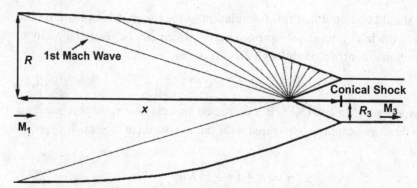

Figure 5.6. A simplified conical Busemann inlet designed for Mach 7 and CR = 6 (Billig et al., 1999). The isentropic waves coalesce and terminate with a conical shock of half-angle θ_s. Assuming inviscid flow, M_3 remains constant at the design point in the constant-area section of the duct.

off-design Mach number flight, the pressure recovery changes, as shown in Fig. 5.7. Here, the pressure recovery is calculated at each axial location with the origin at the shock cone apex and the axial distance normalized by R_3.

The pressure recovery P_3/P_1 is high for the design point, and it remains relatively high even when the inlet operates at off-design Mach numbers, despite the appearance of shock-wave structures in the duct as a result of the departure from isentropic compression conditions. If the inlet has to be

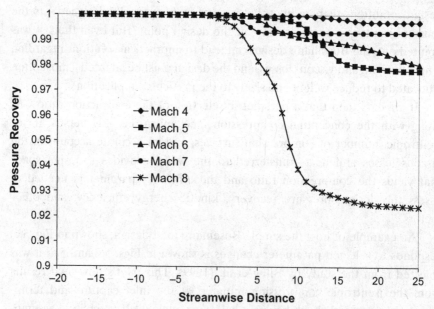

Figure 5.7. The pressure recovery P_3/P_1 remains high even under off-design conditions (after Billig et al., 1999).

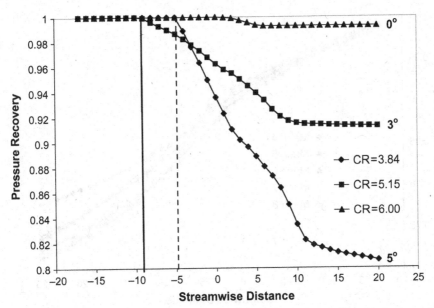

Figure 5.8. The effect of truncation on pressure recovery is considerable because the compression is no longer isentropic.

truncated to facilitate a smaller and lighter structure, the capture ratio can be modified as shown in Fig. 5.8; in this example the truncation is from the nominal CR = 6 to CR = 3.84. An incidence angle appears at the inlet capture, 3° and 5°, as indicated in the figure; the compression is no longer isentropic. The effect of shock-wave structure formation is apparent from the drop in pressure recovery. This has quite a drastic negative effect on the system thrust, but it is partially compensated by the decreased inlet size and weight, and it can be justly judged only when a complete system analysis is done that includes the positive effects of lower weight and reduced viscous and thermal fluid–structure interactions.

It should be noted that the preceding examples result from an inviscid calculation and the viscous effects are strong and may even dominate the loss mechanism in a hypersonic inlet (Billig et al., 1999). As a result, the pressure recovery is considerably lower than the values indicated in Fig. 5.7 and the pressure recovery continuously drops in the duct at the design point to approximately 0.75, as shown in Fig. 5.9. However, because the viscous effects dominate, the difference between on-design and off-design operation is, in turn, diminished.

The Busemann inlet selected in this example may not always be implemented in a mission-driven design. Additional parameters play a role, and they include shock–boundary-layer interactions, wall heat flux, the effect of leading-edge bluntness, integration requirements, and other factors.

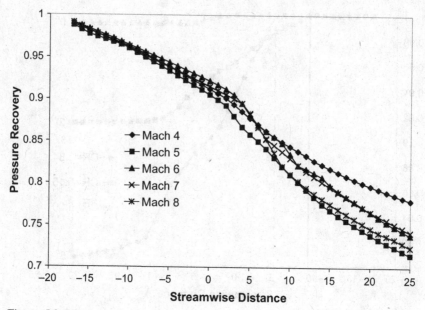

Figure 5.9. Viscous effects on pressure recovery under design and off-design flight conditions.

5.1.3 Flow Interactions and Inlet Design Considerations

5.1.3.1 Inlet Starting

The airflow captured by the inlet largely depends on the vehicle velocity and the engine's ability to pass the mass flow to the nozzle. The minimum area at the inlet throat becomes critical with vehicle acceleration and thus becomes the inlet-mass-flow limiting factor. With supersonic acceleration, the capture is larger than the inlet airstream and the exceeding mass flow is diverted through spillage around the inlet's capture. Heiser and Pratt (1994) describe in detail the flow transition from low speed to high speed in a fixed-geometry inlet with subsonic exit conditions, such as are required for ramjet or turbojet engines. If the throat contraction is severe, or the back pressure increases through other mechanisms, the capture cannot accommodate the flow arriving at the inlet and the spillage increases. The inlet is said to be *unstarted* because of the flow characteristics in the internal duct (Fernandez et al., 2001). The unstart onset is determined by the correlation of the capture Mach number and the contraction ratio for a particular geometry (Van Wie et al., 1996). An unstarted inlet with a subsonic duct flow is characterized by the presence of a normal shock in front of the capture that moves forward as the vehicle accelerates or the back pressure increases to adjust the captured flow to the airstream arriving at the inlet station. For a scramjet inlet, where the flow remains supersonic throughout, the mass flow adjusts through a number of oblique shock waves.

When the boundary layer is thick, as indeed is expected at a hypersonic vehicle inlet station, these shocks interact with the boundary layer, causing separation on the cowl upstream of the capture and resulting in a complex oblique shock- and expansion-wave system following reattachment (Van Wie et al., 1996). An inlet that is unstarted by unfavorable duct or flight conditions may return to a started operation when the unfavorable conditions are removed. Variable geometry, a capture shape designed to facilitate spillage, and a judiciously selected duct bleed may alleviate the unstart problem throughout the flight envelope (Smart and Trexler, 2004).

To determine a boundary of an allowable contraction ratio, the Kantrowitz limit (Kantrowitz and Donaldson, 1945) is widely used because it offers a first-order estimate to permit inlet self-start. This limit is determined for a thermally and calorically perfect gas with the assumption that a normal shock, which appears in the supersonic diffuser, would be pushed upstream toward the throat under back-pressure conditions and would allow the inlet to remain started as long as the normal shock stays within the diffuser; in the limit, the normal shock will be at the throat. If the boundary-layer thickness is neglected and the flow is considered quasi-1D, the contraction ratio in the Kantrowitz limit is

$$
\frac{A_2}{A_{\text{throat}}} = \frac{1}{M_2} \left[\frac{(\gamma+1)\,M_2^2}{(\gamma-1)\,M_2^2+2} \right]^{\frac{\gamma}{\gamma-1}} \left[\frac{\gamma+1}{2\gamma M_2^2-(\gamma-1)} \right]^{\frac{1}{\gamma-1}}
$$

$$
\times \left[\frac{1+\left(\frac{\gamma+1}{2}\right)M_2^2}{\frac{\gamma+1}{2}} \right]^{\frac{\gamma+1}{2(\gamma-1)}}. \tag{5.14}
$$

Further simplifying the inlet operation, if the flow is assumed to remain isentropic throughout the compression process, the area ratio deriving from continuity provides another limit:

$$
\left(\frac{A_0}{A_{\text{throat}}} \right)_{\text{isentropic}} = \frac{1}{M_0} \left[\frac{2}{\gamma+1} \left(1 + \frac{\gamma-1}{2}M_0^2 \right) \right]^{\frac{\gamma+1}{2(\gamma-1)}}. \tag{5.15}
$$

These functions indicate that the contraction ratio, i.e., the inverse of the functions given by Eqs. (5.14) and (5.15), increases with an increasing Mach number. In the isentropic case this increase is without bounds as the high-pressure ratio across the inlet increases continuously, in the absence of any loss mechanism, with the flight Mach number. In the Kantrowitz limit, the allowable contraction reaches a limit that depends on γ:

$$
\lim_{M_2 \to \infty} \frac{A_{\text{throat}}}{A_2} = \frac{(2\gamma)^{\frac{1}{\gamma-1}}(\gamma-1)^{\frac{1}{2}}}{(\gamma+1)^{\frac{\gamma+1}{2(\gamma-1)}}}. \tag{5.16}
$$

Assuming constant γ, this limit is approximately 0.6, and it is approached quickly beyond Mach 6.

Practical inlets operate between these two limits, and the ability to pass the arriving mass flow through the inlet throat depends on a number of factors including boundary-layer thickness, momentum distortion at the throat cross section, and the use of an inlet bleed. The contraction ratio can decrease beyond the Kantrowitz limit for high flight Mach numbers because the shock structure is formed of oblique shocks, thereby generating less loss than would a normal shock assumed at the throat. For example, Smart and Trexler (2004) found through experiments that their inlet remained started at M = 4.68 with a throat-to-capture contraction ratio of 0.465 whereas the Kantrowitz limit indicated 0.653. The trend increases with the Mach number as shown by the experiments collected by Van Wie (2000).

5.1.3.2 Viscous Interactions

High-speed boundary layers developing on the vehicle forebody arrive at the inlet with a high temperature and considerable thickness resulting from the low density and increased viscosity. Whether fully or partially diverted from the inlet's capture, these high-speed boundary layers continue to grow inside the inlet. The effective inlet cross-sectional area is reduced and the heat transferred to the walls and the friction increase. This large flow displacement that is due to thick boundary layers causes the pressure in the inviscid part of the flow to rise, unlike the case of low-speed boundary layers (Anderson, 1989). These effects of the viscous layer on the inviscid flow are known as *viscous interactions*.

If the axial boundary-layer growth is assumed to grow inversely with the Reynolds number as in the laminar flat-plate analysis, namely,

$$\frac{\delta}{x} \sim \frac{1}{\sqrt{\text{Re}}}, \tag{5.17}$$

given that the temperature varies substantially within the boundary layer, the density and the viscosity dependence with the temperature modify the Reynolds number in the viscous layer to a nonnegligible degree. Thus, when temperature effects are considered (Anderson, 1989), the axial boundary-layer growth becomes strongly dependent on the flight Mach number:

$$\frac{\delta}{x} \sim \frac{M_0^2}{\sqrt{\text{Re}}}. \tag{5.18}$$

The pressure rise in the inviscid-flow region leads to a curvature of the streamlines, and, although it tends to reduce the boundary-layer thickness, it cannot compensate for the growing effect that is due to the Mach number. This pressure rise is more severe close to the leading edge and so are the friction coefficient and the heat transfer rates because the rate of boundary-layer growth, $d\delta/dx$, is larger. Therefore this condition is known as the *strong viscous interaction* as opposed to the *weaker viscous interaction* encountered farther downstream when the boundary-layer displacement growth is more moderate. A similarity parameter that describes the laminar viscous interactions along a flat plate is $\bar{\chi}$, defined as

$$\bar{\chi} = M_0^3 \sqrt{\frac{C}{Re}} \tag{5.19}$$

with the parameter C deriving from the differences between the fluid properties at the wall and at the boundary-layer edge:

$$C = \frac{(\rho\mu)_{wall}}{(\rho\mu)_{edge}}. \tag{5.20}$$

Using the hypersonic similarity parameter $K = M\theta$, where θ is the flow deflection across an oblique shock, which for large Mach numbers becomes close to the local-slope angle, the derivation of the induced pressure in the inviscid region located at the boundary-layer edge (Anderson, 1989) depends on the hypersonic similarity parameter K as

$$\frac{p_{edge}}{p_0} = 1 + \frac{\gamma(\gamma+1)}{4}K^2 + \gamma K^2 \sqrt{\left(\frac{\gamma+1}{4}\right)^2 + \frac{1}{K^2}}. \tag{5.21}$$

The dependence of the hypersonic similarity parameter K on the local slope, which is dictated by the boundary-layer growth, $d\delta/dx$, relates it to the viscous-interaction similarity parameter $\bar{\chi}$ as

$$K^2 = M_0^2 \left(\frac{d\delta}{dx}\right)^2 \sim \frac{M_0^2}{\sqrt{Re}}\sqrt{C} \equiv \bar{\chi}. \tag{5.22}$$

For large similarity parameters, $K \gg 1$, the interaction is *strong*, and the effect on the induced pressure scales is

$$\frac{p_{edge}}{p_0} = 1 + a_1\bar{\chi}, \tag{5.23}$$

and for weak interactions, $K \ll 1$, the interaction is considered weak, and the effect is described by

$$\frac{p_{edge}}{p_0} = 1 + b_1\bar{\chi} + b_2\bar{\chi}^2. \tag{5.24}$$

For air with $\gamma = 1.4$ and an insulated flat plate, the solution offered by Hayes and Probstein (1959) leads to

$$\textit{strong interactions,} \quad \frac{p_{edge}}{p_0} = 0.514\bar{\chi} + 0.759; \tag{5.25}$$

$$\textit{weak interactions,} \quad \frac{p_{edge}}{p_0} = 1 + 0.31\bar{\chi} + 0.05\bar{\chi}^2, \tag{5.26}$$

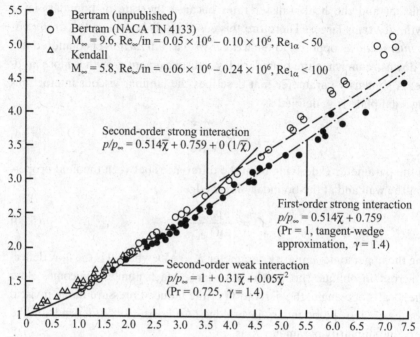

Figure 5.10. Correlation of inviscid induced pressure on a flat-plate prediction with experimental data (Hayes and Probstein, 1959).

with a value of $\bar{\chi} \simeq 3$ marking the separation between the weak- and the strong-interaction regimes. The data provided by Hayes and Probstein (1959), shown in Fig. 5.10, indicate that Eqs. (5.25) and (5.26) correlate reasonably with the experimental data. The theoretical prediction begins to depart from the experimental data at higher $\bar{\chi}$ as the Reynolds number becomes smaller, for example, in the vicinity of the leading edge, where the Knudsen number decreases and the continuum assumption, from which the correlations made in Eqs. (5.25) and (5.26) were derived, becomes inaccurate.

The boundary-layer growth in the inlet, as a result of the interactions just described, is the effective reduction in the capture area, which can be quite severe, particularly if the boundary layer formed along the vehicle's forebody is not entirely diverted. Goonko and Mazhul (2002) indicate that a "flow-rate factor," defined as the ratio of the free-stream flow captured area to the physical inlet area A_∞ / A_0, is as low as 60% if a diverter is not present.

As the boundary layer develops along the inlet walls it transitions from the laminar conditions assumed in this analysis to turbulent conditions. Although more resistant to adverse pressure gradients, the turbulent boundary layer also increases the local rates of heat transfer (Ault and Van Wie, 1994). A particular design has to balance these effects for the selected flight conditions.

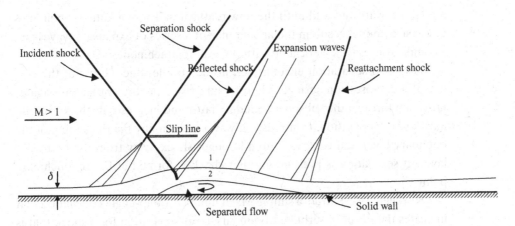

Figure 5.11. Planar shock-wave/boundary-layer interaction.

5.1.3.3 Shock–Boundary-Layer Interactions

Controlling the inlet flow depends to a large extent on the ability to control the shock-wave structure at the capture and within the inlet duct. In turn, the inlet capture shock-wave structure is dictated by the vehicle's forebody shape and the engine flow rate. Within the inlet the shock-wave–boundary-layer interactions play a significant role. When sufficiently strong, these shock waves impinge on the boundary layers that are sensitized by adverse pressure gradients caused by a pressure raise in the combustion chamber, leading to flow separations and producing several adverse effects on the inlet operation. Furthermore, the local boundary-layer distortion generates a new structure of shock waves and modifies the inlet-flow structure. It is a situation that leads to pressure losses, reduced effective flow cross section, localized high thermal loads, potential flow unsteadiness, and the danger of increased upstream interactions through the separated region. A simplified description of the shock-wave-induced boundary-layer separation is shown in Fig. 5.11 (after Shapiro, 1953). The impinging shock penetrates the boundary layer to the sonic line. The associated pressure rise is transmitted upstream in the subsonic region and causes a local boundary-layer thickening; under the adverse pressure gradient, a region of flow separation appears. The thickening of the boundary layer upstream of the impinging shock causes a local bending of the streamlines and the onset of oblique shock waves that eventually coalesce into a stronger shock. This shock intersects the impinging shock and generates a slip line. The part of the impinging shock that penetrates the boundary layer is bent because the Mach number changes continuously – in region 2 in the figure – and generates a series of weak waves that coalesce into a stronger shock. In the region downstream of the incident shock, the pressure outside the boundary layer – region 1 – is larger than that inside the layer – region 2 – and the streamlines are bent toward the wall until the flow reattaches. These additional bendings of the streamlines result in the appearance of a series of expansion waves and a compression shock associated with the wall reattachment.

A typical separation encountered inside an inlet duct occurs at the corner of a compression surface. The pressure rise near the corner propagates upstream through the subsonic boundary layer and increases its thickness and may lead to separation. Compression waves accompany the thickening of the boundary layer and coalesce into a stronger shock away from the boundary layer. If separation is present, another shock is formed at the reattachment point.

Korkegi (1975) offers a simple Mach number–ramp-angle correlation that indicates the onset of turbulent boundary-layer separation for a shock that is skewed with respect to the ramp:

$$M\theta = 0.3, \tag{5.27}$$

where θ (in radians) is the angle formed by the compressing ramp with the incoming flow. This correlation is based on experimental data with $Re_\delta > 10^5$ and $M < 3.5$. For higher Mach numbers, experiments summarized by Korkegi (1975) provide simple correlations in terms of pressure rise across the corner shock, both for straight – and therefore strictly 2D – and skewed shocks, as follows:

$$\frac{p_2}{p_1} = 1 + 0.3M^2 \quad \text{for} \quad M \le 4.5, \tag{5.28}$$

$$\frac{p_2}{p_1} = 0.17M^{2.5} \quad \text{for} \quad M > 4.5. \tag{5.29}$$

These correlations, shown graphically in Fig. 5.12, are independent of the Reynolds number and wall temperature, which is assumed to have a negligible effect for a first-order estimate as in these equations. Pressure ratios above these correlations lead to corner separation.

A more complex 3D interaction appears in corners of 2D inlets or along the intersection of a vertical fin placed in the inlet. A computational-based description of the complex flow field resulting from this interaction is described by Gaitonde et al. (2001) and is shown in Fig. 5.13. Either or both of the vertical and the horizontal surfaces can contribute to fluid compression. A separation occurs along the line marked S_5 along the vertical wall and a reattachment line, marked R_5, follows downstream. The darker surface marked Surface 1 in the picture spatially follows the streamlines above this separation on the vertical surface. A vortical structure appears inside the separation as a result of the interaction of the fluid from the vertical and the horizontal surfaces. This vortical structure prevents reattachment in the corner, and therefore it leads

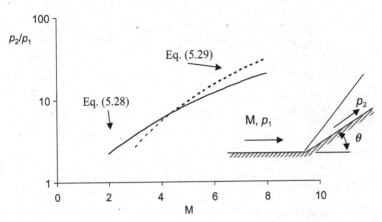

Figure 5.12. Incipient separation pressure rise (after correlations by Korkegi, 1975).

to a region of intense heat load. The boundary layer that has separated on the vertical surface does not reattach as in the simple, 1D separation. Instead, fluid from the corner vicinity arrives at the vertical surface-reattachment location. This is a situation schematically indicated by the darker band in the picture noted as Surface 2. The interaction between the vertical and horizontal layers generates the "saddle" point marked F_1 in the figure and a vortex appears within the vertical wall separation.

The vortical structure that appears at the saddle point F_1 is a 3D reflection of the less evident vorticity induced by the solid boundaries and the shock structure. The presence of the separation on the vertical surface is a favorable region in which this vortex expands and is lifted off the surface, continuing in a vertical direction and away from the plate direction. Streamline S_{12} marks the horizontal surface projection of the S_1 and S_2 intersection, and streamline S_4 marks the edge of the reattachment region.

Figure 5.13. A 2D shock–boundary-layer interaction. Considerable complexity is noticed over the flat-plate-type separation with fluid from one surface (horizontal in this case) entrained in the separate region formed on the other (vertical in the figure).

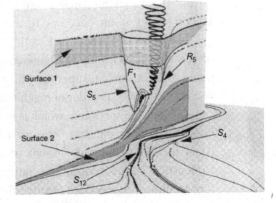

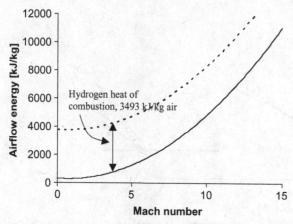

Figure 5.14. Air energy and hydrogen heat of combustion.

This complicated 3D pattern was also noticed in the experiment of Alvi and Settles (1992), in which visualization was used to detect the footprint pattern at the separation location in a plane perpendicular to both the horizontal plate and a vertical, swept-back fin, corresponding to Surfaces 1 and 2 in Fig. 5.13. The lifted vortex separation was evident in the experiment, and a "jet-impingement" effect on the surface, at the reattachment location, was made evident by high surface pressure and increased skin friction.

5.1.4 Advanced Concepts for Inlet-Flow Control

5.1.4.1 Intake Air Energy Management

Air arriving at the scramjet inlet possesses high kinetic energy that quickly becomes comparable with and then larger than the chemical energy in the fuel. Vanderkerckhove and Barrère's (1993) analysis indicates that, as soon as Mach 3 flight is achieved, the air kinetic energy becomes equal to its sensible energy. Figure 5.14 shows the rapid air energy increase with flight speed along a path with a dynamic pressure of 70 kPa. Soon after Mach 8 is reached, the kinetic energy in air becomes 10 times larger than the sensible energy, and before Mach 9 is reached the kinetic energy becomes equal to hydrogen's heat of combustion. This energy balance is remarkable, in particular when compared with the current supersonic flight – under Mach 3 – when the same heat of combustion is one order of magnitude larger than the incoming air's energy.

Through compression, which is needed to reduce the velocity at the combustion chamber entrance, the air temperature becomes high. This leaves room for only a limited amount of additional energy contributed through combustion before severe dissociation prevents deposition of heat in the air flow to ultimately produce positive thrust. The cycle, under these conditions, becomes inefficient and therefore has led to efforts to develop means to

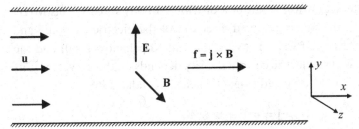

Figure 5.15. Electromagnetic force in a conductive airflow.

manage the engine energy to improve performance. An immediate goal would be to remove the energy from the inlet and transfer it to other components or to carry the energy directly into the nozzle flow, thereby bypassing the combustion chamber. In the nozzle, the expansion can maintain the gas temperature at reasonable levels to avoid dissociation. The opportunity also exists to use part of the energy removed from the inlet for auxiliary systems' operation.

One example of inlet energy management is the KLIN^{TM} cycle (Balempin et al., 2002) described earlier in Chap. 4 along with other cycles. Here, the fuel, at cryogenic conditions, is used to cool the inlet air, thus increasing the air density through heat extraction. A variation of this cycle, called a "deep-cooled turbojet" (DCTJ), uses part of the cryogenic oxygen stored on board by directly injecting it into the airflow (Balempin, 1997). The mixture results in a lower temperature, and therefore it can extend the turbojet operation to higher flight Mach numbers. At the same time, the additional oxidizer allows the injection of more fuel in the combustion chamber than that based on air stoichiometry alone. A similar concept is used in the ATREX design (Takagi et al., 1997) in which the cryogenic fuel is used to cool the intake air of a combined-cycle air-turbo ramjet. The low-speed flight range is supported by a turbojet, which, with cooled intake and a fan-assisted ramjet, has extended operation to Mach 5.3. The LACE concept, also described in Chap. 4, or the liquid–air-collection rocket–ramjet engine (LACRRE), similarly uses cryogenic fuel to cool the intake air; however, the main goal of these concepts is to liquefy a certain amount of air for use as the oxidizer during the engine's rocket component operation (Escher, 1997; Macaron and Surmanov, 1997).

Energy extraction from the inlet has additional advantages that accompany the increased air density: The compression relies less on area changes, thereby reducing wall friction and heat transfer and offering the ability to implement a design that is less sensitive to flight-condition changes.

5.1.4.2. Flow Deceleration Using a Magnetic Field

A body force can be exerted on the inlet's airflow through application of a magnetic field once the airflow has been ionized. If a flow, such as that shown schematically in Fig. 5.15, is subject to both an electric field **E** and a magnetic field **B**, their interaction will result in a distributed body force **f**. This MHD force depends on the magnitude and direction of the electric current **j** and the magnetic field **B**, as $\mathbf{f} = \mathbf{j} \times \mathbf{B}$ (Vatazhin and Kopchenov, 2000) and the gas conductivity σ. The electric current density depends on the air velocity, its conductivity, and the electric and magnetic fields, as (Jahn, 1968)

$$\mathbf{j} = \sigma(\mathbf{E} + \mathbf{u} \times \mathbf{B}). \tag{5.30}$$

The energy associated with the MHD effect, **jE**, can be either extracted or added to a flow, depending on the sign of the **jE** product: $\mathbf{jE} < 0$ will promote energy extraction whereas with $\mathbf{jE} > 0$ energy would be added to the flow field, leading to acceleration.

Energy extraction from the hypersonic inlet is clearly the desired goal. This energy, in the form of an electric field, can be used directly to support auxiliary systems or applied through another MHD process to accelerate the flow in the nozzle and contribute to thrust production. It is thus an extremely attractive means to transfer inlet kinetic energy to the nozzle flow, bypassing the combustion chamber; it also provides a means to control the amount of energy transferred between components (Lichford et al., 2000). The way to achieve MHD control, however, is not a simple technological achievement. Large magnets are necessary to produce appreciable body forces, and for the air ionization to become electrically conductive is a complex process in itself; once ionized, the particles' motion is influenced by the magnetic-field orientation, and through momentum transfer the entire flow field is accelerated or decelerated.

A simple case analysis is helpful to evaluate the extent of the estimated performance of a MHD decelerator. If the flow is assumed to be calorically and thermally perfect, 1D, inviscid, and without internal heat generation, under steady-state conditions and with a negligible magnetic field resulting from the flow of ionized particles, the equations of motion reduce to

$$\rho U = \text{const},$$
$$\rho U \frac{dU}{dx} = -\frac{dp}{dx} + jB, \tag{5.31}$$
$$\rho U \frac{d}{dx}\left(c_p T + \frac{U^2}{2}\right) = jE.$$

The equation of state is added to these, along with the specifications for the electric-field current and the gas conductivity:

$$p = \rho RT,$$
$$j = \sigma(E - UB), \tag{5.32}$$
$$\sigma = \sigma(\rho, T).$$

Further, assuming that the electric and magnetic fields $E(x)$ and $B(x)$ are known, the system can be solved in a closed form. As an example (Jahn, 1968), if an isothermal flow is assumed, i.e., $dT/dx = 0$, combining the momentum and the energy equations leads to

$$\rho U \frac{dU}{dx} = \frac{E}{B}\left(\rho U \frac{dU}{dx} + \frac{dp}{dx}\right).$$ (5.33)

Using the equation of state and continuity, the pressure gradient is substituted by

$$\frac{dp}{dx} = RT \frac{d\rho}{dx} = -\rho U RT \frac{1}{U^2} \frac{dU}{dx},$$ (5.34)

which provides a relation between the velocity and the ratio of the electric field to the magnetic field:

$$\frac{E}{B} = \frac{U^3}{U^2 - a^2_{\text{isothermal}}},$$ (5.35)

where the quantity RT has been substituted with $a^2_{\text{isothermal}}$, the isothermal speed of sound. This relation has a singularity when the velocity approaches the speed of sound. The sign indicates that the electric or the magnetic fields have opposite effects on subsonic or supersonic flows.

If the result in Eq. (5.35) and the electric-field current are further substituted into the energy equation, an expression that provides a solution for the velocity is obtained as

$$(\rho U)U^3 \frac{dU}{dx} = \sigma E^2 a^2,$$ (5.36)

where a indicates the isothermal speed of sound and the quantity ρU is a constant. A simple solution is obtained for σE^2 assumed constant as

$$U^* = \left[1 + 4 \frac{\sigma E^2 L}{(\rho U)U_0^2}\left(\frac{a}{U_0}\right)^2 x^*\right]^{1/4}$$ (5.37)

where $U^* = U/U_0$ is the velocity normalized with the channel entrance velocity and $x^* = x/L$ is the distance normalized with the channel length L. Other solutions are obtained if other assumptions are made regarding the electric or magnetic fields when energy equation (5.36) is integrated (Jahn, 1968).

A similarity parameter appears in Eq. (5.37) in the form of $\sigma E^2 L/(\rho U)U_0^2$; it represents the ratio of magnetic body forces to inertial forces, and it thus indicates the extent of the electromagnetic interaction. Similar parameters are obtained in the solution to Eq. (5.37) when other assumptions are made. All these solutions indicate that this interaction is rather weak, as shown in Fig. 5.16. Here the flow is assumed to be only slightly supersonic; the effect

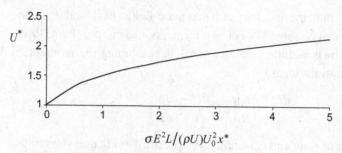

Figure 5.16. Effect of nondimensional magnetic body force on a 1D isothermal flow in a constant-area channel.

can be strengthened if the isothermal assumption is relaxed or if the area is allowed to change.

The analysis developed in the previous pages has been quite restrictive, and, indeed, other assumptions would lead to slightly different results. The adiabatic-flow assumption could be invoked instead of the isothermic assumption made here; however, it requires an infinite air electric conductivity to preclude flow heating. If the limit of a very large Mach number is studied, the solution offered by Vatazhin and Kopchenov (2000) is then recovered. This case indicates that considerable aerodynamic losses are induced in the flow, and, in fact, these losses grow faster than the flow deceleration.

If any thermodynamic path restriction is relaxed, both the electric and the magnetic fields can be prescribed and a relatively complex solution can be obtained (Jahn, 1968). Then the velocity and Mach number equations can be written as

$$\frac{dU}{dx} = \frac{\sigma B^2}{p} \frac{1}{1 - M^2} (U - \xi)(U - \eta), \tag{5.38}$$

$$\frac{dM}{dx} = \frac{\sigma B^2}{pa} \left\{ \frac{1 + [(\gamma - 1)/2] M^2}{1 - M^2} \right\} (U - \xi)(U - \zeta), \tag{5.39}$$

where several characteristic velocities appear that are derived from the magnitudes of the electric and the magnetic fields:

$$\xi = \frac{E}{B},$$

$$\eta = \frac{\gamma - 1}{\gamma} \xi, \tag{5.40}$$

$$\zeta = \frac{1 + \gamma M^2}{2 + (\gamma - 1) M^2} \eta.$$

It is of particular interest to note the dependence of the velocity U and Mach number derivative signs on the electric and magnetic fields as described by

Table 5.1. *Dependence of velocity and Mach number gradients on the characteristic velocities given in Eqs. (5.40) (Resler and Sears, 1958)*

M	U	dU/dx	dM/dx
< 1	$< \zeta$	+	+
	$\zeta < U < \eta$	+	−
	$\eta < U < \xi$	−	−
	$> \xi$	+	+
> 1	$< \eta$	−	−
	$\eta < U < \xi$	+	−
	$\zeta < U < \xi$	+	+
	$> \xi$	−	−

these characteristic velocities. Table 5.1 (after Resler and Sears, 1958) illustrates this dependence. Here both subsonic and supersonic flows are included. Deceleration of the supersonic flow exists as long as the characteristic velocities η and ξ are larger than the flow velocity U. The U–M diagram shown in Fig. 5.17 (after Resler and Sears, 1958) indicates the domains to which the electric and magnetic fields must be constrained to result in flow deceleration. Here the lines of $U = \eta$ and $U = \xi$ are shown along with the function $\zeta(M)$ in the supersonic range. Not all values are attainable in this diagram. Choking, for example, can be achieved, in principle, only at the singular points $U = \eta$ and $U = \xi$ under the restrictive assumptions of this analysis. But for deceleration to exist, the velocity U must be confined to regions a and d. Only in these regions do both the velocity and the Mach derivatives have a negative slope. The implication of the existence of these domains is that deceleration of the supersonic flow in the inlet will occur for only a judicious selection of the electric and the magnetic fields to maintain the velocity $U < \frac{\gamma-1}{\gamma}\frac{E}{B}$ or $U > E/B$.

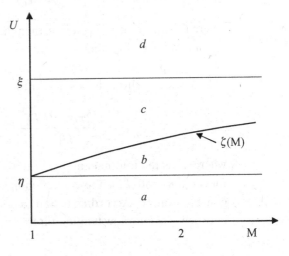

Figure 5.17. Domains of velocity and Mach numbers in the supersonic range (after Resler and Sears, 1958).

When more constraints imposed in this analysis are removed, the area can be allowed to vary and the electric and magnetic fields can be optimized axially, along the inlet duct; considerable improvement in the flow deceleration can thus be achieved. However, beyond these improvements a significant parameter that dictates the efficiency of the MHD process remains the air conductivity, which dictates the magnitude of the electric current density and hence the body force that appears in the equations of motion, Eqs. (5.31). Under normal conditions this parameter is not particularly large; it therefore deserves some attention.

The air conductivity is determined by

$$\sigma = \frac{q^2 n_q}{m_q v_c},\tag{5.41}$$

where q is the particle charge, n_q is the charged-particle concentration, m_q is the charged-particle mass, and v_c is a collision frequency that depends on the "thermal velocity" corrected by the collisional effects:

$$v_c = \sqrt{\frac{8kT}{\pi m_q}} \left(\sum_i n_i Q_i \right),\tag{5.42}$$

where n_i and Q_i are the particle concentration and collisional cross section, respectively and account, in this equation, for both electron and ion collisions.

If the ionization process is assumed to follow Saha's development for monoatomic gases (described by Jahn, 1968) and extended later to diatomic gases (as in the example by Vatazhin and Kopchenov, 2000), according to which the process is characterized by a single, reversible reaction for a pure gas,

$$A + \varepsilon_i \rightleftarrows A^+ + e,\tag{5.43}$$

where ε_i is the ionization energy, then the activated-particle concentration can be found by assuming thermodynamic equilibrium with activated-particle temperature equal to the gas temperature. Hence the analysis produces the degree of ionization α through Saha's relation,

$$\frac{\alpha^2}{1 - \alpha^2} = \frac{2 (2\pi m)^{\frac{3}{2}} (kT)^{\frac{5}{2}}}{p h^3} \left(\frac{f_+^i}{f_A^i} \right) \exp\left(-\frac{\varepsilon_i}{kT} \right),\tag{5.44}$$

where f_+^i is the internal energy modes' contribution to the partition function for the ions and f_A^i is the corresponding partition function of the gas undergoing ionization. According to Saha's relation, the degree of ionization ranges between $\alpha = 0$ for low temperatures, $T \to 0$, and a value of $\alpha = 1$ for $T \to \infty$.

Table 5.2. *Example of 1D calculation of flow deceleration with* $M_e = 1.3$ *and* $B/UE = 0.4$ *(Vatazhin and Kopchenov, 2000)*

M_0	P_e/P_0	U_e/U_0	T_e/T_0
5	7.2	0.484	3.5
10	25.5	0.431	11.0

Ion concentration and hence the gas conductivity σ increase proportionally with the degree of ionization α.

The fraction of ionized air atoms is not particularly large, and it reaches only slightly more than 1% when the air temperature exceeds 6000 K (Hansen, 1976). Figure 5.18 shows the degree of ionization of a nitrogen species at 1 atm. The electron and N^+ are negligible below 1 atm unless temperatures become quite elevated.

An example of expected deceleration of hypersonic flow through energy extraction is given in Table 5.2, using the calculation by Vatazhin and Kopchenov (2000). Here the relation between the magnetic and the electric field was chosen as $B/UE = 0.4$ and the exit Mach number was chosen fixed at $M_e = 1.3$. For both cases the velocity was reduced to approximately 43%–48% of the initial value, which is quite substantial; the corresponding temperature increase is also included and, for the case of Mach 10 at the channel entrance, it indicates quite a large value at the end of compression.

The method of flow deceleration through energy bypass by use of MHD may have undesired effects. The temperature gradient in the boundary layer creates a gradient in the air electrical conductivity, leading to variations in the

Figure 5.18. Nitrogen ionization at high temperatures.

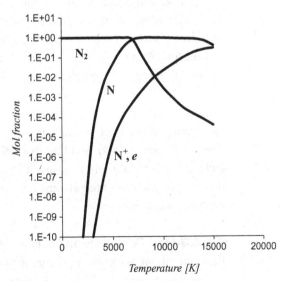

way energy is extracted from adjacent layers, and, by affecting the body forces, it influences the fluid dynamic interactions. The presence of axial forces that is due to magnetic- and electric-field interactions,

$$f_x = \sigma B(E - UB), \qquad (5.45)$$

can induce boundary-layer separation. Solutions of theoretical analyses offered by Vatazhin and Kopchenov (2000) clearly indicate the separation of the boundary layer subjected to MHD interactions.

The previous analysis of MHD application to a scramjet inlet was developed around the idea of energy extraction from the incoming flow for later deposition in the propulsive nozzle. Yet MHD energy depositions have found additional applications in controlling the flow upstream of the inlet so that optimal operation close to the shock-on-lip condition can be guaranteed for a range of flight conditions with fixed inlet geometries. Macheret et al. (2004) show how selective energy deposition upstream of the inlet's capture can, for example, reduce spillage at flight conditions below the design Mach number. The energy in this case is recovered entirely in the form of heat added to a carefully selected region of the flow, causing a deflection of the streamlines upstream of the inlet. A *virtual cowl* is thus formed in a way that increases the mass capture. The calculations by Macheret et al. (2004) indicate that a 2%–3.5% energy addition to the airstream enthalpy can increase the inlet mass capture by as much as 11%. Following this concept, Schneider et al. (2004) suggest extending MHD application by the addition of energy into the inlet when the flight takes place at Mach numbers higher than the design numbers to avoid ingestion of the shock system, thereby returning to shock-on-lip operation. For this, a short MHD device would be installed on the vehicle forebody, a structural simplification allowing high beam current densities.

MHD application to scramjet inlets is clearly attractive for a number of reasons, some of which were already indicated. Practical implementation remains problematic, mostly because of the large-sized magnets it requires. In the middle range of hypersonic flight, a certain degree of air ionization will be present, facilitating the MHD interaction in the generator. But in the lower flight regime, some of the extracted energy would have to be used to ionize the air arriving at the inlet. Potentially, large pressure losses may occur, as calculated by Riggins (2004), and therefore dictate against the application of MHD techniques in the scramjet engine. The added weight and technical complexities must be also weighted against the potential benefits before the technology becomes feasible.

5.1.4.3 Flow Control Using Fuel Injection

Efficient mixing is clearly a precursor to efficient combustion in high-velocity, air-breathing engines. It is important for hydrogen-fueled engines, and it becomes even more significant for gaseous and liquid hydrocarbon-fueled engines for which chemical kinetic time scales are longer and additional time is required for liquid vaporization. The inlet length can be used to improve mixing if some, or, in the limit, all of the fuel is injected upstream, on the vehicle's forebody or in the inlet. The idea of inlet fuel injection and combustor integration was considered in early scramjet engine designs: Henry and Anderson (1973) suggested injecting a large part of the fuel from in-stream struts downstream of the inlet throat well before the combustor entrance.

This injection solution requires closer inlet–combustor integration and a careful placement of the fuel injectors to minimize the danger of preignition that may be caused by flame propagation upstream through the inlet boundary layer; the potential benefit, however, is great, offering a high degree of fuel-air mixedness at the combustor entrance. Additionally, fuel injection in the inlet would contribute to airflow compression, which is normally accomplished on the inlet compression surfaces; it would also preheat the fuel by using the energy in the air (Vinogradov, 1997). Furthermore, when liquid fuels are used, precombustor fuel injection would enhance the secondary breakup of fuel droplets that is due to interactions with the inlet's shock compression system (Vasiliev et al., 1994). The benefits of properly selected inlet fuel injection can thus be considerable.

In the *shcramjet* concept (Sislian and Dudebout, 1993; Sislian and Parent, 2004; Schwaitzentruber et al., 2005), which involves a shock-induced temperature rise sufficient to ignite the mixture, the fuel is injected from the vehicle forebody or inlet ramp. Computational analyses by Sislian and his collaborators, applied to hydrogen injection through cantilevered struts installed on the inlet ramp at Mach 11, indicated that fuel preinjection increased significantly the engine thrust but the inlet losses were also increased by as much as 40%–120%. This level of pressure loss is excessive; better means of fuel injection would clearly be needed.

In the case of fuel injection in the inlet or farther upstream, a complex but, at the same time, more flexible system is obtained. Here, the air–fuel interactions occur over the entire inlet–isolator–combustor system. Despite the increased complexity, the optimization of this system could result in multiple advantages including (i) mixing enhancement; (ii) shorter isolator and combustor and, consequently, reduced weight and cooling loads; (iii) a more flexible fuel control system because of the possibility of distributing the fuel between the preinjection system and the fuel supplied directly to the combustor; and (iv) the possibility of injecting combinations of liquid and gaseous fuels through different sets of injectors.

A solution for efficient fuel injection and mixing in the inlet was offered by Vinogradov and Prudnikov (1993) in the form of the pylon-based injectors shown in Fig. 5.19.

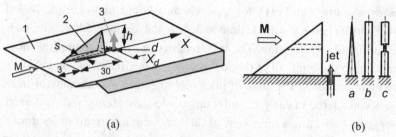

(a)   (b)

Figure 5.19. (a) Jet, 3, injected behind thin pylon, 2, transverse to wall, 1, in supersonic flow; (b) diagrams of possible pylon configurations: *a*, triangular; *b*, rectangular; *c*, rectangular with side-wall releases to enhance aerodynamic jet breakup.

This concept involves thin, swept pylons with the fuel injected in the separated region behind it, transverse to the local flow. The pylon's thickness is maintained at 1.5–2 times the orifice diameter d_{inj}, with a swept leading edge and different cross sections. The pylon's length depends on the material's conductivity and the amount of convective cooling available at its base. The pylons' cross sections can be of different shapes, as shown in Fig. 5.19(b) including (i) triangular, (ii) rectangular, or (iii) a rectangular cross section with side-wall releases for enhanced aerodynamic jet breakup; the latter type is preferable for liquid fuels with increased density and viscosity. These pylons do not cause large pressure losses.

The penetration increase with these pylons is substantial. The study by Golubev and Yagodkin (1979) quoted in the review by Vinogradov et al. (2007) led to a fivefold-to-sevenfold increase over the measured penetration at the same dynamic pressure ratio in the absence of pylons. Figure 5.20 shows the breakup of a liquid jet behind two of these types of injector. Because the fuel is injected in the low pressure behind the pylon, the jet penetrates to the top of the pylon, where it is abruptly turned downstream by the oncoming flow. In essence, the behavior is similar to that of an axial injector from a strut; by

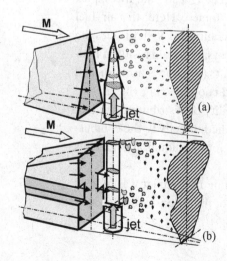

Figure 5.20. A jet injected in the low pressure behind the pylon penetrates to the top of the pylon, where it is abruptly turned downstream by the oncoming flow. Shaping the pylon sides increases the surface area of aerodynamic jet breakup.

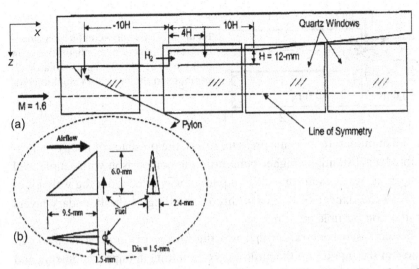

Figure 5.21. A 2D layout of the study by Owens et al. (2001). A thin triangular pylon is located 10 steps upstream of the rearward-facing step where the combustion chamber starts. A pilot hydrogen flame behind the step is used to ignite ethylene or JP-10. Flashback is not encountered in any of the cases in which the isolator flow is maintained at a supersonic level.

contrast, these pylons are thin and cause minimal pressure losses in the inlet (Livingston et al., 2000).

It is further significant that considerable penetration can be accomplished with relatively low dynamic pressure ratios, i.e., less then unity. In most cases, normal injection from the wall injection requires dynamic pressure ratios of the order of 10–15 (Schetz, 1980).

When the fuel is injected into the inlet, there is a danger of some fuel remaining in the slow boundary layer, potentially causing flame propagation upstream. With careful pylon design, the fuel can penetrate through the boundary layer and eliminate this danger. The following two examples of fuel injection upstream of the combustion chamber show that flashback can be avoided.

The study by Owens et al. (2001) used triangular pylons to inject fuel into the isolator of the model shown in Fig. 5.21. The dynamic pressure ratio was maintained at $\bar{q} = 0.7$ when a liquid fuel (JP-10) was used and at $\bar{q} = 3.0$ for gases (ethylene). Without the pylon, boundary-layer separation was noticed with an evident effect on the strengthening of the isolator's shock system. Although penetration was double at $\bar{q} = 3$ for injection behind the pylon, the effect on the isolator's shock train was insignificant. A similar result was observed in the liquid injection case. Briefly, the study by Owens et al. (2001) indicated the following:

- Thin pylons with sharp leading edges do not introduce significant pressure losses or distortion in the isolator airflow.

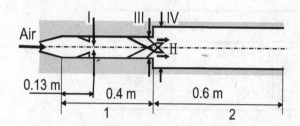

Figure 5.22. Isolator–combustion chamber and injection configurations in Shikhman et al. (2001): 1, isolator; 2, combustion chamber; I–IV, fuel-injector locations.

- Even at moderate dynamic pressure ratios, the presence of the pylon promoted a substantially higher penetration in comparison with simple wall injection. As a result, the entire liquid jet was lifted from the wall, eliminating the danger of flashback through seeding of the boundary layers with a combustible mixture.
- The increased penetration that was due to the pylon presence brought most of the liquid into the airflow core, resulting in improved mixing and creating the possibility of reduced combustion chamber length; the combustion chamber far field showed larger rates of heat release and thus improved combustion efficiency through better mixing.
- Airflow choking with substantial effects on the isolator shock train was not observed for injected equivalence ratios as high as 0.5.

A more complex study described by Shikhman et al. (2001) showed how upstream pylon injectors could be combined with strutlike and other injection options to optimize efficiency. The geometry is shown in Fig. 5.22. This configuration included two sections: the isolator with a slightly divergent area from 40 to 50 mm in height followed by the main fuel combustion zone starting at 80 mm and separated from the isolator by rearward-facing steps. Here, I–IV indicate the availability of injection locations. Shikhman et al. (2001) used methane as the fuel at equivalence ratios $\phi = 0$–0.85 and fuel temperatures of $T_f = 550$–880 K in a Mach 2 airflow.

The fuel distribution varied among various injectors. The combustion efficiencies of five injection combinations are shown in Fig. 5.23. The results indicate the following conclusions:

- High mixing efficiency was achieved within 0.6 m of the duct length with combustion efficiency η_c increasing even as the fuel rate was increased.
- The least effective fuel-injection configuration, from a mixing point of view, was found when the entire fuel flow was injected from wall location III without pylons present; yet this configuration also exhibited a relatively high combustion efficiency when used in combination with injection from pylons II.
- Analyses of heat fluxes and combustion efficiency distributions along the duct indicated that there was no combustion within the insulator in any of these cases.

Case	φ_I	φ_{II}	φ_{III}	Pylons II present
1	1	0	0	Yes
2	0.8	0.08	0.12	No
3	1	0	0	No
4	0.8	0	0.12	Yes
5	0.4	0	0.6	No

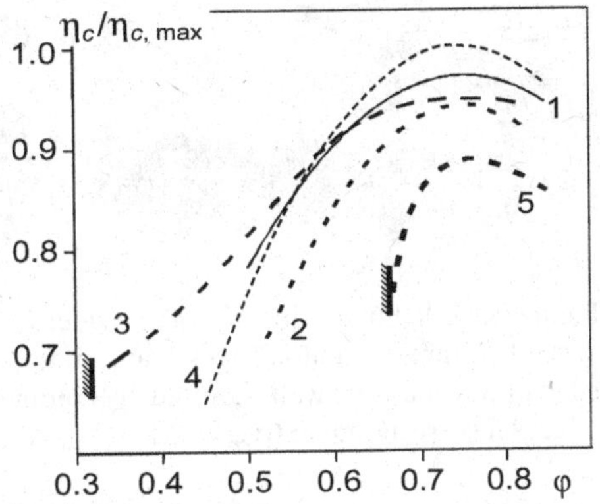

Figure 5.23. Five combinations of injection show increased combustion efficiency when pylons are present.

These examples of upstream injection and combustion showed clearly that flashback can be avoided; they give confidence to the option of fuel injection in the inlet. If incorporated into the engine design, this option must occur without fuel mass losses and with minimal impact on the inlet pressure losses. Additionally, the fuel-injection solution must avoid the need for large supply pressures. The study by Livingston et al. (2000) provides an example of liquid injected into an inlet.

A single ramp was selected along with two thin struts placed at equal distance slightly after the leading edge. The model was placed in a $M_\infty = 3.5$ airstream, and it used as the injectant a mixture of 50% by volume commercial ethylene glycol in water, resulting in a viscosity and surface tension similar to those of JP-10. This mixture was injected from the inlet's wall, as shown in Fig. 5.24, through round orifices transverse to the ramp behind each of two swept pylons of triangular cross section. It was found that the inlet flow field remained started with only minor separated regions when liquid injection rates were used that corresponded to fuel–air equivalence ratios for JP-10 of $\phi \leq 0.45$ and practically all the liquid was lifted into the airflow core. The injectant liftoff from the surface is evident in the Schlieren images shown in Fig. 5.25 for cases without injection [Fig. 5.25(a)] and with liquid injection [Fig. 5.25(b)]. The shock structure remained approximately constant, and only

Figure 5.24. Inlet photograph indicating the pylons' locations on the compression surface: 1, 10° wedge; 2, cowl; 3, pylons; 4, side walls; 5, static pressure taps.

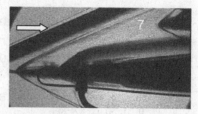

Figure 5.25. Schlieren images of the inlet flow field at M = 3.5 and α = 5°: 1, pylon; 2, cowl; 3, ramp; 4, initial shock wave; 5, upper boundary of side bleed slot; 6, leading edge of the inlet side wall; 7, liquid-fuel plume: (a) without injection and (b) with liquid injection corresponding to φ = 0.22.

at liquid rates corresponding to $\phi > 0.35$–0.4 was a shock curvature observed near the cowl and the boundary layer separated. Larger rates of injectant led to inlet unstart. Once the inlet unstarted, the liquid spread to occupy the entire inlet cross section.

The presence of pylons, absent liquid injection, had a negligible effect on total pressure losses in the inlet; Fig. 5.26 indicates that the measured pressure recovery at the inlet exit had about the same values in both cases. Similar results were observed for $0°$ and $5°$ angles of attack. A pressure loss of 15% was noted at $\phi = 0.3$ because of the momentum exchange between the high-speed airflow and the liquid. This momentum exchange is expected to accelerate the droplet breakup and enhance mixing, potentially compensating for the inlet pressure loss; it may even result in an increased efficiency of the entire system's inlet–combustion chamber. When the inlet unstart was observed, the stagnation pressure loss was close to 50%.

Finally, it is worth noting the experiment by Ogorodnikov et al. (1999), which involved an inlet-equipped combustor operating in dual mode. The fuel was injected from the inlet's cone with additional ports in the combustion chamber. Figure 5.27 shows a diagram of this model. It is a 250-mm-diameter axisymmetric dual-mode scramjet model at $M_\infty = 6.2$ with air total temperatures and pressures of $T_t = 1400$–1600 K and $P_t = 4.8$–5.5 MPa, respectively. Liquid kerosene at rates corresponding to $\phi = 0.4$–0.45 was injected transverse to the flow from the inlet cone at 20 mm from the apex through eight orifices of 0.4-mm diameter. Ignition and flameholding were ensured through

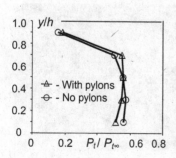

Figure 5.26. Inlet figure indicating the pylons' location on the compression surface: 1, 10° wedge; 2, cowl; 3, pylons; 4, side walls; 5, wall pressure taps.

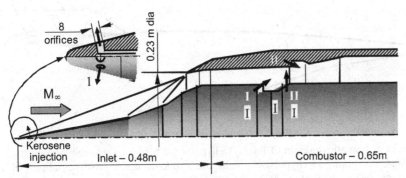

Figure 5.27. Schematic diagram of kerosene injected into an axisymmetric scramjet model: I, II, III, hydrogen injector sets, $T_f = 300$ K; 1, liquid kerosene jets. The first cone angle was 18°, followed by two additional cones of 5° each.

gaseous hydrogen injection at the leading edge of the cavity flameholders on both the inner and outer walls before the entrance into an annular combustor duct. The quantity of hydrogen injected was small, $\phi_{H_2} = 0.15 - 0.2$, the minimal amount experimentally determined as necessary for stable operation of the flameholder. Significantly, under all conditions, stable combustion was achieved without flashback through the boundary layers.

5.1.5 Summary

The inlet is required to exhibit high performance, ensure engine compatibility over the entire operational range, and adjust for effective flow control in response to speed, altitude, and engine airflow changes. With a flight envelope significantly larger than any other air-breathing engine, the scramjet imposes particularly difficult demands on the engine design. Some form of variable geometry is expected, but optimization for all flight conditions based on geometry changes alone will require new and ingenious approaches. Among the major sources of concern is boundary-layer development and stability under elevated temperature. Control of the boundary layer to maintain operability and high performance will most likely continue to impose significant design challenges. Toward that goal, development of accurate, predictive tools for high-enthalpy flows remains a major goal because experiments that covered all flight conditions would be prohibitively expensive.

A larger degree of integration of the inlet with the engine thermodynamic cycle is expected for hypersonic flight than for the more conventional flight regimes. The considerable energy present in the airstream arriving at the inlet's capture is both an advantage and a challenge. The presence of cryogenic fuels in some applications can be used to densify and store some of the intake air for operation of the rockets in the combined cycles. However, the

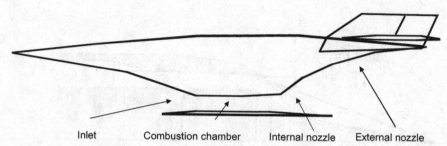

Inlet   Combustion chamber   Internal nozzle   External nozzle

Figure 5.28. Schematic diagram of the X-43 hypersonic vehicle. To accommodate the large expansion required in hypersonic flight, the nozzle occupies a substantial part of the afterbody (after McClinton, 2002).

large air energy content limits the amount of additional energy that can be provided by the fuel for vehicle acceleration. Bypassing part of this energy from the inlet to the nozzle would be quite attractive if a MHD process could be made efficient in terms of the weight of the additional components required. Despite the general desire to extract energy from the inlet, the airflow control for high performance is of such basic importance that MHD processes that in fact deposit energy in the inlet are considered for their ability to control the shock-wave structure and the mass-flow capture.

Fuel preinjection in inlets or isolators holds considerable potential to enhance mixing, flame stability, and combustion efficiency for scramjet engines. The use of thin pylons leads to significant improvements without incurring large penalties in the form of pressure loss and shock generation. This additional distribution of fuel creates an additional "knob" to control the fuel distribution to adapt the engine to the flight envelope and to improve inlet–engine compatibility. Significant issues remain to be solved before implementation in practical devices, including the effects of changing the flow structure in the inlet as a result of angles of attack and side-slip excursions, engine transients, and integration with the airframe.

5.2 Nozzles

At the end of the combustion process, the air enthalpy has increased sufficiently to generate thrust through expansion in the nozzle. Potential energy is traded for flow acceleration until, ideally, the nozzle exit pressure equals that of the atmosphere at the flight altitude. Given that the hypersonic vehicle operates with a large nozzle pressure ratio, this expansion is extensive and requires long nozzles. It is expected therefore that the scramjet nozzle would be of an *open* type, with much of the vehicle's lower surface acting as the part of the nozzle similar to the configuration exhibited by the X-43 experimental vehicle shown schematically in Fig. 5.28 (after McClinton, 2002). Part of the expansion takes place in an internal nozzle, after which a considerable part of the vehicle's afterbody constitutes the external part of the nozzle.

Because a substantial part of the vehicle is dedicated to nozzle expansion, considerable lift and pitch moments are produced by the pressure distribution on this part of the afterbody, complicating the nozzle design and vehicle integration. During hypersonic flight, the engine thrust is only slightly larger than the vehicle's drag; hence the efficiency of the expansion process and the thrust angle relative to the flight direction become critical for the vehicle's flight dynamics. Optimization over the entire flight range and possibly over all the propulsion modes for combined cycles further increase the nozzle design complexities.

Nozzle efficiency is affected by a number of factors that are due to the design itself and to the flow characteristics at the beginning of the expansion. Wall friction in the nozzle is one major source of loss, and it suggests the adoption of a short design with substantial divergence. As much as 2% of the ideal thrust – defined as the maximum thrust achievable from an energetic point of view if the expansion were completed to the altitude pressure – can be lost because of friction along the nozzle walls (Anderson et al., 2000); most of these losses occur in the first part of the nozzle where the pressures are higher. The divergence is another potential source of loss because the flow departs from the ideally perfect axial orientation. Often the nozzle must be shorter than the ideal expansion would require; the flow then remains underexpanded and some thrust is consequently lost. The gain in nozzle efficiency in these cases is considered unjustified when the additional weight required is considered.

In addition to the losses associated directly with the nozzle design, additional thrust losses are caused by the flow structure at the nozzle entrance. These flow nonuniformities are caused by a number of factors that trace their source to the inlet efficiency and flow distortion at the end of compression, the fuel injection in the combustion chamber configuration, mixing and combustion efficiency, boundary-layer development, and interaction with the shock-wave structure. All these factors are design dependent and are subject to configuration optimization and integration.

A review of nozzle sources of inefficiencies cannot be complete without referring to the composition of the hot gases at the combustion chamber exit entering the nozzle. If, given the high temperatures, combustion is incomplete at the nozzle entrance and dissociated species are present, the rapid expansion in the nozzle may lead to freezing of the flow, eliminating the chance for a recombination reaction of these dissociated radicals and therefore leading to energy loss. Although this is not a particularly significant problem for moderately supersonic flight conditions, this becomes a serious concern for the hypersonic flight regime when the temperature increase through compression in the inlet may leave little room for the addition of heat through combustion before the temperature reaches levels when a great deal of dissociation is present.

All the factors previously listed lead to quite restrictive nozzle designs that are greatly dependent on the thermoaerodynamics of the other components, the vehicle configuration, and the flight dynamics.

REFERENCES

Alvi, F. S. and Settles, G. S. (1992). "Physical model of the swept shock wave/boundary-layer interaction flowfield," *AIAA J.* **30**, 2252–2258.

Anderson, G. Y., McClinton, C. R., and Weidner, J. P. (2000). "Scramjet performance," in *Scramjet Propulsion* (E. T. Curran and S. N. B. Murthy, eds.), Vol. 189 of Progress in Astronautics and Aeronautics, AIAA, pp. 369–446.

Anderson, J. D. (1989). *Hypersonic and High Temperature Gas Dynamics*, McGraw-Hill.

Andrews, E. H., Jr. and Mackley, E. A. (1976). "Analysis of experimental results of the inlet for the NASA Hypersonic Research Engine Aerothermodynamic Integration Model," NASA TM X-3365.

Andrews, E. H., Jr., Russell, J. W., Mackley, E. A., and Simmonds, A. L. (1974). "An inlet analysis for the NASA Hypersonic Research Engine Aerothermodynamic Integration Model," NASA TM X-3038.

Ault, D. A. and Van Wie, D. M. (1994). "Experimental and computational results for the external flowfield of a scramjet inlet," *J. Propul. Power* **10**, 533–539.

Balempin, V. V. (1997). "Combined cycle for SSTO rocket: Definition of key technologies," in *Proceedings of the 13th International Symposium on Air Breathing Engines*, AIAA, pp. 987–994.

Balempin, V. V., Liston, G. L., and Moszée, R. H. (2002). "Combined cycle engines with inlet conditioning," AIAA Paper 2002–5148.

Bertram, M. H. (1958). "Boundary layer displacement effects in air at Mach numbers 6.8 and 9.6," NACA TN 4133.

Billig, F. S., Baurle, R. A., Tam, C.-J., and Wornom, S. F. (1999). "Design and analysis of streamline traced hypersonic inlets," AIAA Paper 99–4974.

Escher, W. J. D. (1997). "Cryogenic hydrogen-induced air liquefaction technologies for combined-cycle propulsion applications," in *The Synerget Engine*, SAE Progress in Technology Series, PT-54, Society of Automotive Engineers, pp. 163–174.

Fernandez, R., Trefny, C. J., Thomas, S. C., and Bulman, M. J. (2001). "Parametric data from a wind tunnel test on a rocket-based combined-cycle engine inlet," NASA TM-2001–107181.

Gaitonde, D. V., Visbal, M. R., Shang, J. S., Zheltovodov, A. A., and Maksimov, A. I. (2001). "Sidewall interaction in an asymmetric simulated scramjet inlet configuration," *J. Propul. Power* **17**, 579–584.

Goonko, Y. P. and Mazhul, I. I. (2002). "Some factors of hypersonic inlet/airplane interactions," *J. Aircr.* **39**, 37–50.

Goldfeld, M. A., Starov, A. V., and Vinogradov, V. A. (2001). "Experimental study of scramjet module," *J. Propul. Power* **17**, 1222–1226.

Golubev, A. G. and Yagodkin, V. I. (1979). "Methods of dispersity and liquid spray droplet concentration definition by using of light diffusion integral characteristics," *Science Notes of CIAM*, No. 867 (in Russian).

Hansen, C. F. (1976). "Molecular physics of equilibrium gases – A handbook for engineers," NASA SP-3096.

Hayes, W. D. and Probstein, R. F. (1959). *Hypersonic Flow Theory*, Academic Press.

Heiser, W. H. and Pratt, D. T. (1994). *Hypersonic Airbreathing Propulsion*, AIAA Education Series.

Henry, J. R. and Anderson, G. Y. (1973). "Design considerations for the airframe-integrated scramjet," NASA TM X-2895.

Holland, S. D. and Perkins, J. N. (1990). "Mach 6 testing of two generic three-dimensional sidewall compression scramjet inlets in tetrafluoromethane," AIAA Paper 90-0530.

Hsia, Y.-C., Gross, B., and Ortwerth, J. P. (1991). "Inviscid analysis of a dual mode scramjet inlet," *J. Propul. Power* **7**, 1030–1035.

Jahn, R. G. (1968). *Physics of Electrical Propulsion*, McGraw-Hill.

Kantrowitz, A. and Donaldson, C. duP. (1945). "Preliminary investigation of supersonic diffusers," NACA WR L-713.

Korkegi, R. H. (1975). "Comparison of shock-induced two- and three-dimensional incipient turbulent separation," *AIAA J.* **13**, 534–535.

Lichford, R. J., Bityurin, V. A., and Lineberry, J. T. (2000). "Thermodynamic cycle analysis of magnetohydrodynamic-bypass hypersonic airbreathing engines," NASA TP-2000–210387.

Livingston, T., Segal, C., Schindler, M., and Vinogradov, V. A. (2000). "Penetration and spreading of liquid jets in an external–internal compression inlet," *AIAA J.* **38**, 989–994.

Macaron, V. S. and Surmanov, V. N. (1997). "Rocket-ramjet engine of air liquefaction cycle (LACRRE): Performance analysis and experimental investigations," in *Proceedings of the 13th International Symposium on Air Breathing Engines*, Vol. 2, AIAA, pp. 1259–1265.

Macheret, S. O., Schneider, M. N., and Miles, R. B. (2004). "Scramjet inlet control by off-body energy addition: A virtual cowl," *J. Propul. Power* **42**, 2294–2302.

Mahoney, J. J. (1990). *Inlets for Supersonic Missiles* (J. S. Przemieniecki, series editor-in-chief), AIAA Educational Series.

McClinton, C. R. (2002). "Hypersonic technology...Past, present and future," presented at a seminar at the University of Florida. Available from C. Segal.

Ogorodnikov, D. A., Vinogradov, V. A., Petrov, M. D., Shikhman, Yu. M., and Stepanov, V. A. (1999). "Some results of CIAM scramjet technology," AIAA Paper 1999–4030.

Ortwerth, P. J. (2000). "Scramjet flowpath integration," in *Scramjet Propulsion* (E. T. Curran and S. N. B. Murthy, eds.), Vol. 189 of Progress in Astronautics and Aeronautics, AIAA, pp. 1105–1293.

Owens, M. G., Mullagiri, S., Segal, C., and Vinogradov, V. A. (2001). "Effects of fuel preinjection on mixing in a Mach 1.6 airflow," *J. Propul. Power* **17**, 605–610.

Resler, E. L., Jr. and Sears, W. R. (1958). "The prospects for magnetoaerodynamics," *J. Aeronaut. Sci.* **25**, 235–245.

Riggins, D. W. (2004). "Analysis of the magnetohydrodynamic energy bypass engine for high-speed airbreathing propulsion," *J. Propul. Power* **20**, 779–792.

Schetz, J. A. (1980). "Turbulent flowfield mixing and injection process," in (Martin Summerfield, ed.), Vol. 68 of *Progress in Astronautics and Aeronautics*, AIAA.

Schneider, M. N., Macheret, S. O., and Miles, R. B. (2004). "Analysis of magnetohy-drodynamic control of scramjet inlets," *J. Propul. Power* **42**, 2303–2310.

Schwaitzentruber, T. E., Sislian, J. R., and Parent, B. (2005). "Suppression of premature ignition in the premixed inlet flow of a shcramjet," *J. Propul. Power* **21**, 87–94.

Shapiro, A. H. (1953). *The Dynamics and Thermodynamics of Compressible Fluid Flow*, Wiley, Vol. 1.

Shikhman, Yu. M., Vinogradov, V. A., Yanovskiy, L. S., Stepanov, V. A., Shlyakotin, V. E., and Pen'kov, S. N. (2001). "The demonstrator of technologies – Endothermic hydrocarbon fueled dual mode scramjet," AIAA Paper 2001–1787.

Sislian, J. P. and Dudebout, R. (1993). "Hypersonic shock-induced combustion ramjet performance analysis," in *Proceedings of the 11th International Symposium on Air-Breathing Engines*, AIAA, pp. 413–420.

Sislian, J. P. and Parent, B. (2004). "Hypervelocity fuel/air mixing in a scramjet inlet," *J. Propul. Power* **20**, 267–272.

Smart, M. K. and Trexler, C. A. (2004). "Mach 4 performance of hypersonic inlet with rectangular-to-elliptical shape transition," *J. Propul. Power* **20**, 228–293.

Takagi, Y., Monji, T., Tomike, J., Tanatsugu, N., Sato, T., and Kobayashi, H. (1997). "Development study of air intake for ATREX engine," in *Proceedings of the 13th International Symposium on Air Breathing Engines*, AIAA, Vol. 1, pp. 204–211.

Vanderkerckhove, J. and Barrère, M. (1993). "Energy management," in *Proceedings of the 11th International Symposium on Air Breathing Engines*, AIAA.

Van Wie, D. M. (2000). "Scramjet inlets," in *Scramjet Propulsion* (E. T. Curran and S. N. B. Murthy, eds.), Vol. 189 of Progress in Astronautics and Aeronautics, AIAA, pp. 447–511.

Van Wie, D. M., Kwok, F. T., and Walsh, R. F. (1996). "Starting characteristics of supersonic inlets," AIAA Paper 1996–2914.

Van Wie, D. M. and Molder, S. (1992). "Applications of Busemann inlet designs for flight at hypersonic speeds," AIAA Paper 92–1210.

Vasiliev, V., Zakotenko, S. N., Krasheninnikov, S. Yu., and Stepanov, V. A. (1994). "Numerical investigation of mixing and augmentation behind oblique shock waves," *AIAA J.* **32**, 311–316.

Vatazhin, A. B. and Kopchenov, V. I. (2000). "Problem of hypersonic flow deceleration by magnetic field," in *Scramjet Propulsion* (E. T. Curran and S. N. B. Murthy, eds.), Vol. 189 of Progress in Astronautics and Aeronautics, AIAA, pp. 891–938.

Vinogradov, V. (1997). "Review of Russian hydrocarbon fueled scramjet technology," presented at the workshop on Russian Hypersonic Technologies, APL-JHU.

Vinogradov, V. A. and Prudnikov, A. G. (1993). "Injection of liquid into the strut shadow at supersonic velocities," SAE-931455, SAE Aerospace Atlantic Conference, Society of Automotive Engineers.

Vinogradov, V. A., Shikhman, Yu. M., and Segal, C. (2007). "A review of fuel pre-injection in supersonic, chemically reacting, flows," *ASME Appl. Mech. Rev.* **60**, 139–148.

6 Supersonic Combustion Processes

6.1 Introduction

With the broad range of flying conditions in the hypersonic regime, the processes in the supersonic combustion chamber are subject to large variations in thermodynamic conditions. At the low range of the hypersonic flight regime, the heat deposition in the combustion chamber is relatively large compared with the incoming flow energy; hence the heat deposition substantially reduces the air speed and a large pressure rise is experienced with possible flow separations. At the higher range of the hypersonic regime, close to Mach 25, which is considered the upper envelope of air-breathing propulsion, the heat addition may amount to only 10% of the incoming airflow enthalpy. The heat-release effects are less pronounced. The airspeed in the combustion chamber itself may be hypersonic, and the heat deposition is distributed over a longer distance following mixing and chemical reactions; the pressure rise associated with combustion is less pronounced and is mostly due to the internal geometry of the combustion chamber.

The aerothermodynamic processes in the supersonic combustion chamber are complex and closely related. The comparable time scales lead to a closely coupled turbulent mixing and chemical reaction rates. Combustion cannot be initiated until mixing has been achieved at a molecular level, and, in turn, in regions where combustion has taken place, the temperature rise and the chemical composition changes modify the parameters responsible for mixing. This close coupling cannot be, in general, separated in the supersonic combustion chamber. Associated with the turbulent mixing and finite-rate chemical reactions are significant 3D flow features. In particular, in the high-speed regime when shock waves may be present and subsonic flow regions may be embedded in the generally supersonic flow, real-gas effects must be included and large temperature and gas composition gradients exist. The complexities of the scramjet engine can seldom be studied through simplified analyses that would separate the major processes involved. Yet a substantial amount of understanding has been achieved through independent studies of turbulent mixing and combustion and the factors that influence these processes. The following discussion refers to turbulent mixing associated with molecular mixing, which is the precursor to initiating chemical reactions. Mixing under both conditions of parallel and transverse or angled fuel–air systems is reviewed, and several practical solutions are offered to enhance and accelerate mixing. Combustion with particular emphasis on finite-rate chemical kinetics processes is described next. The issue of flameholding is particularly acute in a high-speed flow and, as of yet, is not understood sufficiently so designers can offer reliable solutions for the entire flight regime; some relevant issues related to flameholding are therefore included. Finally, fuel candidates and fuel management in the scramjet engine are reviewed.

6.2 Time Scales

There are large differences between the time scales of the physical and chemical processes encountered in a supersonic combustion ramjet. Assuming a vehicle flying at the low end of the hypersonic regime, $M = 6$–8, the combustor entrance Mach number is expected to be in the range $M = 2$–3; thus, in a reasonably long engine of no longer than several meters, the fluid residence time is of the order of a few milliseconds. This is the time available for all aerothermal processes to be completed efficiently so that heat can be extracted from the fuel and thrust can be produced in the engine's nozzle. Gaseous fuels must penetrate into the airstream following injection and mix to a molecular level so that molecular collisions can result in chemical reactions and heat release; additional processes exist in which condensed-phase fuels are used, including jet breakup and droplet vaporization.

A suggested generic time-scale comparison is offered by Warnatz et al. (1996) and is reproduced in Fig. 6.1. Depending on temperature, pressure, composition, and concentration, chemical reactions in general can span over a vast time scale range from 10^{-10} more than 1 s. Molecular transport processes cover a smaller range between 10^{-4} and 10^{-2} s. The fast chemical processes correspond to equilibrium conditions and the long chemical processes to frozen conditions. In these cases chemical processes can be uncoupled from the flow-field analysis; however, these situations seldom exist in the scramjet engine.

From a chemical kinetics point of view, the time required to complete the exothermic reactions results from the addition of the *ignition delay time* and the *burning time*, i.e., the heat-release time. The separate treatment and summation of the ignition delay time and the burning time is appropriate in the distributed-reaction regime (Balakrishnan and Williams, 1994), which is characteristic for high Reynolds numbers and moderate Damköhler numbers.

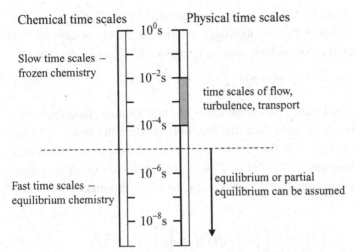

Figure 6.1. Time scales in chemically reacting flows (Warnatz et al., 1996).

Here, the Damkhöler number can be appropriately defined based on the time associated with the Kolmogorov time and the reaction time τ_c as $\mathrm{Da}_k = \tau_k/\tau_c$ (Williams, 1985). The supersonic combustion conditions fall in a region of Da_k around 1.

The definition of the ignition delay time has received different formulations as either the characteristic time inferred from the reaction rate of a particular branching reaction (Balakrishnan and Williams, 1994), as the time required by the mixture to attain 5% of the equilibrium temperature (Rogers and Schexnayder, 1981) or as the time of maximum rate of increase of temperature (Colket and Spadaccini, 2001). If a hydrogen-fueled propulsion system is considered, the specific reaction-rate constant k (defined in Chap. 2) of the $H + O_2 \rightarrow OH + H$ chain-branching reaction is of particular interest and propagates with a rate of $k[H][O_2]$, where [H] and $[O_2]$ are the molar concentrations of atomic hydrogen and molecular oxygen, respectively. The ignition delay time based on this reaction is therefore given by

$$t_{\mathrm{ign}} = 1/k[O_2] \quad \text{with} \quad k = 3.52 \times 10^{16} \, T^{-0.7} \exp(-8580/T), \qquad (6.1)$$

where the constants in the specific reaction rate are given by Balakrishnan and Williams (1994) from the study by Masten et al. (1990). The underlying assumption when Eq. (6.1) is used is that ignition is disconnected from mixing and therefore the transport phenomena do not play a significant role during the ignition process. From operational conditions ranging between flight Mach numbers of 6 and 8, Mitani et al. (2001) calculated the ignition delay time of this reaction as an order of magnitude of 10^{-5} s. These values correspond to stagnation temperatures in the range of 1200–1400 K and clearly increase rapidly at lower temperatures.

The burning time is defined as the time needed to achieve 95% of the equilibrium temperature. For the hydrogen–air system, this time depends on pressure and initial stagnation temperature (Rogers and Schexnayder, 1981):

$$t_r = 3.25 \times 10^{-4} P_b^{-1.6} \exp\left(-0.8 \cdot T_{st}/1000\right). \tag{6.2}$$

The pressure used in Eq. (6.2) is a local value and may change along the burning region with an appreciable quantity, particularly when the flame is distributed over a larger zone, as is the case for the turbulent flames in a supersonic combustion environment. Mitani et al. (2001) suggest the use of an averaged pressure for a selected zone that depends on the pressure exponent in Eq. (6.2) as follows:

$$P_b = \frac{1}{x_l} \left\{ \int_{x_1}^{x_2} [P_w(x)]^n dx \right\}^{1/n}, \tag{6.3}$$

where P_w is the wall pressure distribution, x_l is the axial distance for averaging, and n is the pressure exponent in combustion time Eq. (6.2). Under the same conditions at which the ignition delay time was evaluated before, the combustion time is of the order of 10^{-3} s, and is therefore two orders of magnitude slower than the ignition delay time.

Therefore, assuming that mixing at a molecular level has been achieved, combustion time estimates for a hydrogen–air system are comparable with the residence time for a hypersonic vehicle in the Mach 6–8 regime and, unless a flameholding mechanism is in place to extend the residence time, the exothermic chemical reactions cannot be completed within the combustion chamber.

Regardless of which mixing mechanism present in the flow is dominant – through shear-layer development in parallel flows by massive momentum and mass exchange as in the case of transverse fuel–air injection (both of which are subsequently discussed) – molecular-level mixing precedes the onset of chemical reactions, and because mixing is the longest process encountered in the supersonic combustion chamber (Ferri, 1973), it becomes the limiting factor in the supersonic combustion chamber.

6.3 Fuel–Air Mixing

The fluid residence time is only of the order of milliseconds in a scramjet engine of reasonable length; therefore the mixing processes are determining factors in the complex ensemble of physical processes that ultimately lead to heat release and thrust generation. In general, fuel injection and mixing in practical devices are complex processes involving turbulent 3D flows accompanied by large velocity gradients with subsonic flow regions embedded in a generally supersonic flow in the presence of shock waves and large chemical composition and temperature variations. In the practical flow field of interest in the scramjet, engine mixing cannot be uncoupled from the effects of heat release and chemical composition changes because the parameters responsible for mixing, including density and the transport properties, are directly affected by the chemical and thermal changes taking place that, in turn, depend on the fuel–air mixing degree. Moreover, the time scales of the processes involved are comparable with respect to each other, further strengthening the coupling between fluid dynamics and chemical kinetics. Despite this close interaction, which makes the description of these flow fields a considerable challenge, significant progress has been made in the analysis and the description of individual processes, starting with simplified mixing models and expanding the analyses with additional features likely to be encountered in practical applications.

All conceivable mixing mechanisms exist in the scramjet flow field, including simple diffusion, mixing of parallel streams of different velocities, densities, and chemical composition, and bulk mixing resulting from non-parallel streamlines, which is accompanied by large vortical structures and considerable momentum exchange. Mixing between parallel flows because of the development of the shear layers at the flow boundaries is a process that may take place with relatively low momentum loss but requires long distances to result in complete, molecular-level mixing. Therefore, in a practical device, this type of mixing would require a long combustion chamber, which leads to weight and heat transfer penalties. Injection of the fuel into the airstream transversly or at a certain angle quickly results in a substantial convection of bulk fuel mass into the airstream and, accompanied by additional 3D effects such as swirling motion, achieves mixing within a shorter distance. This type of mixing results, however, in considerable momentum losses and generation of sometimes strong shock waves. Both cases are met in the supersonic combustion chamber and therefore deserve attention. The simple model of diffusion between two coflowing streams in the absence of a velocity difference, thus with zero shear and under laminar conditions, is not typical for flows in a scramjet engine and is not addressed here. This simplified analysis was included in Chap. 3. Instead, turbulent flow with large differences in velocity between adjacent flow streams is discussed in what follows.

6.3.1 Parallel, Unbounded, Compressible Flows

Parallel flows of dissimilar properties appear in the scramjet engines at the boundary of a recirculation region or of a jet injected at a low angle into the main high-speed airflow. The difference in the streams' velocity can be large; densities and gas composition at the interface are in general also different. Momentum is transferred to the slower flow and mass is exchanged between the two streams at their interface. The turbulent small scales are responsible for the momentum transport ultimately leading to the molecular-level mixing

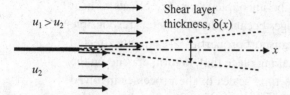

Figure 6.2. Schematic of shear-layer-thickness development.

required for the initiation of chemical reactions. Analysis of high-speed, chemically reacting mixing layers, which are of interest in hypersonic applications, requires an in-depth study that solves all of these small length and time scales. This continues to remain a considerable challenge both for experimental and theoretical studies (Warnatz et al., 1996; Oran and Boris, 2001).

A simplified schematic of the shear-layer development for unbounded flows of different initial velocities is shown in Fig. 6.2. When the two streams, assumed in this case to have negligibly thin boundary layers, coalesce, the differences in velocity, density, and transport properties result in the development of a shear layer in which the two fluids mix. For large relative Reynolds numbers, defined in relation to the characteristic dimensions and velocity of the shear layer as

$$\mathrm{Re} \equiv \frac{\delta \, \Delta u}{\nu}, \tag{6.4}$$

which is based on the relative velocity $\Delta u = u_1 - u_2$, an interface between the streams ensues across which rapid mixing takes place as a result of the entrainment of the two streams into the large fluid turbulent structures developing in the mixing layers. Thus, even when molecular diffusivity has small values, the diffusive flux across this interface can be large (Dimotakis, 1991).

It should be noted, before we evaluate the effect of flow regimes on shear-layer development, $\delta = \delta(x)$, that within shear layers where chemical reactions are present, sublayers can be defined in which (i) mixing is achieved at a molecular level within a layer of thickness δ_{mix} and (ii) chemical reactions take place in a layer of thickness δ_{react}. Thus, as Dimotakis (1991) points out, the extent of the chemically reacting layer, $\delta_{\mathrm{react}} = \delta_{\mathrm{react}}(x)$, can be defined by the product of the fraction of the reacting region within the molecularly mixed flow, $\delta_{\mathrm{react}}/\delta_{\mathrm{mix}}$, the fraction of the molecularly mixed flow within the visual shear layer, $\delta_{\mathrm{mix}}/\delta$, and the development of the shear layer itself, $\delta(x)$. This interpretation is useful for identifying the various time scales representing the processes involved; however, it should not be understood to imply that the mixing or the reacting sublayers develop proportionally to the visual shear layer as the flows progress in downstream. In fact, the mixing layers develop around the large-scale structures that separate the initially different streams; only within these mixing layers can chemical reactions take place.

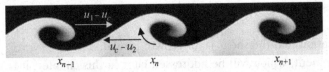

Figure 6.3. Fluid entrainment along the large structures developing in a 2D shear layer. Across the boundary of the large turbulent structures, fluid is entrained from the initially separate streams; diffusion then results in the formation of a layer of molecular-level mixed fluid, identified in the figure by the gray shades, within which a chemical reaction can take place [Atsavapranee and Gharib (1997) using Dimotakis's (1991) schematics and nomenclature].

Figure 6.3 [visualization by Atsavapranee and Gharib (1997) using Dimotakis's (1991) schematics and nomenclature] identifies the large vortical structures in a 2D shear layer that result from the onset of an instability at the coalescence of the two flow streams. Fluid from the two streams is entrained along the regions in which some degree of mixing has occurred and that in turn bound a narrow region of molecularly mixed flow. Within this region chemical reactions can take place and progress at rates that depend on the local thermodynamic conditions. In a reference frame moving with the vortical structures with a velocity u_c, the velocities of the fluids on the two sides of the interface are $u_1 - u_c$ and $u_c - u_2$.

Macroscopic processes, rather than microscopic diffusion, dominate the compressible mixing process. For example, a diffusion increase, which is expected with an increase in the fuel stagnation temperature, does not materialize in purely shear-layer mixing tests (Wendt et al., 1997). Increasing the stagnation temperature of one or both layers prior to mixing results in velocity increases that reduce compressibility and hence shear forces. This reduction in shear force and macroscopic turbulence more than offsets any diffusion increase from increased temperature. The major roles in the development of the shear layer and, implicitly, the mixing process are played by the fluids' velocity, density, and compressibility. These effects are discussed in what follows.

6.3.1.1 The Definition of the Convective Mach Number

The role played by compressibility on the development of the shear layer as the dominant host for the mixing processes is particularly significant for high-speed subsonic and supersonic flows as those of interest in the context of this chapter. It has been observed that the development of the shear-layer thickness $\delta(x)$ is influenced less by density differences between the two streams as it is caused by the velocity gradient (Brown and Roshko, 1974). This suggests the significant role of compressibility in the development of the turbulent shear layer (Papamoschou and Roshko, 1986) along with the other influencing factors such as pressure gradient and heat release if chemical reactions are present.

With the observation that the shear-layer growth is related to large vortical structure development, which will be addressed later in this chapter, it is reasonable to analyze the compressibility effects on the shear-layer growth within the reference frame of the motion of the vortical structures (Bogdanoff, 1983; Papamoschou and Roshko, 1986), which travel with the convective velocity u_c. Thus the relative convective Mach numbers associated with the two streams are

$$M_{c_1} = \frac{u_1 - u_c}{a_1}, \quad M_{c_2} = \frac{u_c - u_2}{a_2}, \tag{6.5}$$

where a_1 and a_2 are the speeds of sound in the two free streams. The convective Mach numbers defined as in Eq. (6.2) describe the effects that compressibility has on the development of the shear layer. If a relation can be found between the two convective Mach numbers defined by Eq. (6.2), the convective velocity u_c can in turn be found. For incompressible flows, such a relation is offered by the assumption that a stagnation point exists between adjacent vortical structures shown in Fig. 6.3 (Dimotakis, 1986). Under these conditions,

$$p_1 + \frac{1}{2}\rho_1(u_1 - u_c)^2 \approx p_2 + \frac{1}{2}\rho_2(u_c - u_2)^2. \tag{6.6}$$

Because the differences in static pressure across the mixing layer are further neglected, the convective velocity can be directly related to velocity and density ratios as follows:

$$\frac{u_1 - u_c}{u_c - u_2} \approx \sqrt{\frac{\rho_2}{\rho_1}}, \tag{6.7}$$

from which the convective velocity is obtained as

$$\frac{u_c}{u_1} \approx \frac{1 + r\sqrt{s}}{1 + \sqrt{s}}, \tag{6.8}$$

where $r \equiv u_2/u_1$ and $s \equiv \rho_2/\rho_1$. A similar result is obtained for compressible flows if the isentropic relation for stagnation pressure is used (Bogdanoff, 1983; Papamoschou and Roshko, 1986):

$$\left(1 + \frac{\gamma_1 - 1}{2}M_{c_1}^2\right)^{\frac{\gamma_1}{\gamma_1 - 1}} \approx \left(1 + \frac{\gamma_2 - 1}{2}M_{c_2}^2\right)^{\frac{\gamma_2}{\gamma_2 - 1}} \tag{6.9}$$

Equations (6.5) along with their definition give a relation between the convective Mach numbers:

$$M_{c_1} = \sqrt{\gamma_1/\gamma_2}M_{c_2}. \tag{6.10}$$

It should be noted that the assumption of a zero pressure difference across the large structures that separate the two streams is satisfied in all cases of subsonic convective Mach numbers because any difference in velocity between the two streams that is due to a possible disturbance would result in the adjustment of the flow to the new conditions. The solution for the convective velocity given by approximation (6.6) was found to be in good agreement with the experimental results for flows of moderate compressibility (Brown and Roshko, 1974).

6.3.1.2 Two-Dimensional Shear-Layer Growth – Velocity and Density Dependence

From dimensional analysis, the shear layer grows proportionally with the free-stream velocity and inversely with the convective velocity (Papamoschou and Roshko, 1986), i.e.,

$$\frac{\delta}{x} \sim \frac{\Delta u}{u_c}. \tag{6.11}$$

Experimental evidence gives a proportionality constant in approximation (6.8) that ranges between 0.16 and 0.18, which is taken by most researchers (Brown and Roshko, 1974; Dimotakis, 1986) in the midrange to obtain a simplified relation,

$$\frac{\delta}{x} = 0.17\frac{(1-r)(1+\sqrt{s})}{1+r\sqrt{s}}, \tag{6.12}$$

where the effect of the density difference between the two free streams is emphasized.

From considerations of fluid entrainment in the large scales developing in the shear layer, Dimotakis (1986) concludes that the cross-sectional area of the shear layer increases between consecutive flow structures as a result of mass addition promoted by the fluid entrainment as

$$\frac{A_n}{t} = \text{const}\left[(u_1 - u_c)(x_{n+1} - x_n) + (u_c - u_2)(x_n - x_{n-1})\right], \tag{6.13}$$

where n is the index of a given vortex in the shear layer, as shown in the schematic in Fig. 6.3 and the time t appears in the equation to select an instant in the temporal development of the shear layer. Substituting for the area, $A_n = 1/2\,\delta_n\,(x_{n+1} - x_{n-1})$, and returning to laboratory coordinates, we find that the prediction for the shear-layer growth becomes

$$\frac{\delta}{x} = \text{const}\frac{(1-r)(1+\sqrt{s})}{1+r\sqrt{s}}\left[1 - \frac{(1-\sqrt{s})/(1+\sqrt{s})}{1+2.9(1+r)/(1-r)}\right], \tag{6.14}$$

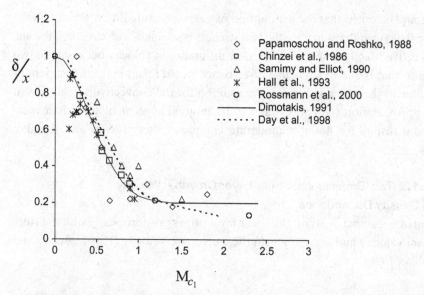

Figure 6.4. Effect of compressibility on the shear-layer growth. The function suggested by Eq. (6.12) matches earlier data, and it is not clear that the shear-layer growth exhibits the asymptotic behavior indicated by this correlation. Experimental results at higher convective Mach numbers seem to indicate a better representation based on linear stability analysis.

which reduces to approximation (6.9) if the second term in the square bracket of (6.14) disappears. The difference between the two derivations resides in the spatial dependence of the shear-layer growth as opposed to the temporal growth treatment in the derivation leading to approximation (6.9). When adjusted to the experimental results of Brown and Roshko (1974), the constant appearing in approximation (6.11) is taken as 0.17 with the observation that the uncertainty in the determination of boundaries of the shear layer may be as large as 20% (Papamoschou and Roshko, 1986) or even more (Dimotakis, 1991). The sources of this uncertainty are attributed not only to experimental inaccuracy but to fundamental fluid physics that has not been included so far in the analyses, including the boundary-layer structure of the two flows leaving the splitter plate (Eggers and Torrence, 1969), turbulence level in the two streams (Schetz, 1980), splitter-plate wake (Bradshaw, 1966), etc.

6.3.1.3 Compressibility Effects on Shear-Layer Growth

Experimental results indicate a sharp reduction in shear-layer growth as compressibility, expressed by the convective Mach number M_{c_1}, increases. Figure 6.4 includes a compilation of results obtained in several studies including the data from Papamoschou and Roshko (1988), Chinzei et al. (1986), Samimy and Elliot (1990), Hall et al. (1993), and Rossmann et al. (2000). The shear-layer growth is shown in the figure normalized by the growth corresponding to the incompressible case, i.e., $M_{c_1} = 0$. Included in the figure is a curve fit suggested by Dimotakis (1991) following the observation that, in the absence of substantial experimental evidence above $M_{c_1} > 2$, the data seem to indicate that an asymptote is reached. This curve fit is described by the function

$$f(M_{c_1}) = 0.2 + 0.8\, e^{-3M_{c_1}^2}. \tag{6.15}$$

Using a linear stability analysis, Day et al. (1998) suggested that the shear-layer growth based on the central-mode analysis varies slightly from the empirical correlation suggested by Dimotakis (1991) in the lower convective Mach number range, up to about 1.5, but allows for a further reduction in shear-layer growth as the convective Mach number continues to rise. This appears to be supported by the limited experimental data available at large convective Mach numbers. Discrepancies can be attributed to limitations of both the experiments and theoretical analyses. The experimental data collected in the studies included in Fig. 6.4 made use of bounded channels in which any wave generated is contained; therefore the bounded facilities tend to feed this energy into the shear layer, enhancing its growth. However, linear stability analyses do not capture the presence of shock waves that are present in these flows.

Clearly the data and the theoretical analyses suggest that compressibility has a strong effect on the shear-layer growth that drops abruptly for even moderate increases in the convective Mach number. The implications are negative for scramjet-type flows if rapid mixing is a requirement and a shear-layer-type mixing is part of the fuel–air mixing mechanism. Other types of mixing mechanisms must be sought.

The results of Hall et al. (1993) at relatively low convective numbers, below 0.3, seem to indicate the existence of an opposite trend, suggesting that the shear-layer growth increases in this region with increasing convective Mach numbers. Because the experimental conditions that led to these results were all collected with gases at low density, it was suggested by Hall et al. (1993) that additional mechanisms may be involved but not accounted for in the analysis of shear-layer growth based on approximations (6.9) and (6.11); the linear stability analysis shown in Fig. 6.4 does not capture this behavior either. The existence of an additional convection velocity that exists simultaneously with the convective velocity of the large structures has been observed experimentally by Rossmann et al. (2000) and raised the possibility of a co-layer structure that modifies the shear-layer growth.

6.3.1.4 Effects of Heat Release on the Shear Layer

When chemical reactions accompany mixing in shear layers, the heat deposition results in lower density and increased volume. Although the shear-layer displacement increases because of dilatation effects, an outward velocity component appears, reducing the entrainment process (Hermanson and Dimotakis, 1989). The vorticity thickness, defined by Brown and Roshko (1974) through a vorticity-thickness parameter $C_\delta = \Delta u/(\partial u/\partial y)_{max}$, decreases, which would result in decreased entrainment that offsets the shear-layer displacement increase. Through another perspective, the reduced entrainment is due to the reduction in the turbulent stress $\tau = \overline{\rho u'v'}$ in the shear layer without encountering significant changes in the velocity correlation $\overline{u'v'}$ (Dimotakis, 1991).

6.3.1.5 Mixing Within the Shear Layer

The large-scale structures seen in Fig. 6.3 create a convoluted interface around which mixing takes place. As a result of the mixing around this interface, the mixing process, particularly molecular-level mixing, takes place at time and spatial scales covering the entire spectrum. Therefore, within the shear layer, the proportion of fluid mixed at the molecular level, in any given cross section, can have values that change both temporally and spatially. Accounting for all the small scales involved in the molecular mixing process remains a substantial challenge for both measurement and theoretical modeling.

It has been noted (Dimotakis, 1991) that the degree of mixing in a turbulent shear layer depends primarily on a local Reynolds number defined, for example, as in Eq. 6.1. The effects of diffusion and viscosity are less significant in the case of the typical scramjet flow because they balance each other in a typical, turbulent, chemically reacting gaseous flow, i.e., the Schmidt number, $Sc \equiv v/D \approx 1$. As compressibility increases, the turbulent structure becomes less organized (Clemens and Mungal, 1990; Samimy et al., 1992) and the accuracy of molecular mixing analysis becomes increasingly uncertain. Yet experimental evidence (Petullo and Dolling, 1993) indicates that as the shear layer develops, it becomes more organized in the downstream direction and clearly is more organized than the boundary layers present at the fluids' coalescence point. That, along with the observation that as the large structures develop they entrain initially unmixed fluids, led to the suggestion that there exists an initial region of unmixedness, and thus an initial *mixing-transition* length. This length, estimated from experiments in incompressible free shear layers to correspond to a Reynolds number defined as in Eq. (6.4) based on shear-layer parameters, is of the order of 10000 (Dimotakis, 1991). At the end of this mixing-transition length, the turbulent structures in the shear layer have evolved to a degree that allows mixing at a molecular level to begin. The exact determination of the mixing-transition length is, of course, complicated by the difficulty of determining when the mixing at small fluid scales begins because it is expected to depend on numerous factors including initial turbulence, compressibility, etc.

If a model that assumes that a layer of molecularly mixed fluids develops within the visual shear layer is adopted, which has a thickness ratio δ_m/δ representing a fraction of the entire mixed layer, the immediate questions are these: (i) What values does this fraction take, and (ii) how is this value changed by the flow parameters, most notably Re? The mixing fraction is clearly related to the fluid concentration at any given axial location where practically all values of concentrations can be expected from 100% fluid from the high-speed flow to 100% fluid from the low-speed flow. A conserved scalar C_s is thus appropriately defined as the fraction of one of the constituents, for example, the high-speed fluid, in the mixture, with values that can range from 1 to 0 when normalized by the concentration in the initially unmixed high-speed flow. In the mixed region one would expect the probability density functions (PDFs) for each stream to be clustered around the region of mean concentration; therefore a normalization of the PDF can be done by accounting for all PDFs around this location (Clemens and Paul, 1995) as

$$\int_{-\infty}^{\infty} P(C_s, \eta) \, dC_s = 1, \qquad (6.16)$$

where the cross-section location η is a nondimensionalized spatial coordinate defined as $\eta = (y - y_{0.5})/\delta$, where y is the cross-section location and $y_{0.5}$ is the location where the mean concentration is equal to 0.5. Thus the probability of the mixed fluid at any location η is defined from the conserved scalar as

$$P_m(\eta) = \int_{\varepsilon}^{1-\varepsilon} P(C_s, \eta) \, dC_s, \qquad (6.17)$$

where ε is a threshold limit that depends on the analysis accuracy, with higher accuracy obtained as ε is reduced. In turn, the mixing fraction is obtained from the summation of this probability throughout the shear-layer cross section as

$$\delta_m/\delta = \int P_m(\eta) \, d\eta. \qquad (6.18)$$

Using this analysis applied to concentration measurements in free-shear layers, Clemens and Paul (1995) found the mixing fraction δ_m/δ within the range of 0.45–0.48 for moderately compressible flows, $M_c = 0.35$–0.82, which appears to be in good agreement with the solution obtained by Dimotakis (1991) under the incompressible assumption. This seems to indicate that there is a small effect of compressibility on the mixing fraction (Mungal et al., 1985), unlike the strong effect compressibility has on the development of the visual shear layer. This leads to the conclusion that compressibility may act only at the large scales without affecting the small-scale flow features at which molecular-level mixing takes place.

6.3.2 Mixing of Angled or Transverse Flows

Rapid mixing requirements in the scramjet engine favor injection of the fuel at large angles relative to the airstream; the strong momentum and mass

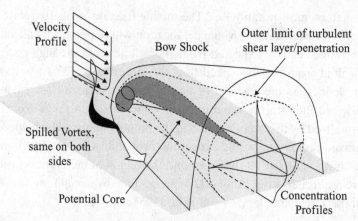

Figure 6.5. Model of a transverse, underexpanded jet in a supersonic airstream. The injected plume forms a barrel shock, generating a bow shock that leads to separation and the formation of a recirculation region in front of the jet. Vortices spill around the barrel shock, which bends in a downstream direction. An additional recirculation region forms at the jet downstream stagnation point. Downstream the injectant angle relative to the supersonic airstream decreases and a turbulent shear layer forms in which mixing continues, facilitated by the streamwise vortices spilling off of the turned injectant plume.

exchange lead to rapid mixing but also lead to nonnegligible viscous losses. Many injection configurations can be conceived, including sonic or supersonic gaseous jets, liquid jets, or dual-phase injectants, with various injection orifice shapes, sizes, orientations, and pressure ratios. For simplicity a schematic of a particular case, represented by a single, underexpanded, transverse jet in a supersonic flow, is shown in Fig. 6.5 (from Portz and Segal, 2004). Significant shock and viscous interactions occur in this case, enhancing mixing but at the same time contributing to pressure losses that reduce engine thrust. The underexpanded jet forms a barrel shock that represents a blockage for the incoming supersonic flow, thus resulting in the generation of a bow shock, which leads to boundary-layer separation and the formation of a recirculation region in front of the jet. Axial vortices spill around the barrel shock, which bends in a downstream direction a short distance after the injection location. An additional recirculation region forms at the jet downstream stagnation point. These recirculation regions may play a significant role in chemically reacting flows because they represent regions of low speed where flames can be sustained. Cool air passing through the oblique shocks curving around the bow shock extends the time required to complete chemical reactions beyond the fluid residence time, and a flame cannot be sustained in these regions although the equivalence ratio may otherwise be within flammability limits. As the ratio of gas-to-air dynamic pressures decreases, the injectant angle relative to the supersonic airstream is reduced and the vortices' axis of rotation aligns more nearly with the airstream. A turbulent shear layer forms, as illustrated in

Fig. 6.5, in which mixing continues, facilitated by the streamwise vortices spilling off of the turned injectant plume.

This model presumes a boundary-layer thickness smaller than the injector diameter so that the gas jet penetrates through the boundary layer, forming the strong bow shock in the supersonic free stream. In the body-integrated scramjet design, considered for many aerospace vehicles, a long inlet ramp with a continuous, strong, adverse pressure gradient is likely to result in a thick boundary layer in the combustor at the fuel-injection location. A high stagnation temperature, intrinsic to hypersonic flight, does not facilitate bleeding this boundary layer, so the simplified assumption of a thin boundary layer at the injection site may not materialize in practice. The extent of the separation in front of the jet and implicitly the position and strength of the bow shock associated with it depend on a number of factors, including the boundary-layer thickness, the jet fluid dynamic characteristics, the local interaction between the airflow and the heat released by means of chemical reactions, and the local wall heat transfer (Ferri, 1973). All these factors result in a complex fluid structure that furthermore is subject to changing conditions by both the supersonic flow upstream and the subsonic regions that may develop downstream of the jet because of heat release.

The jet penetration into the supersonic airstream, usually defined as the distance from the wall reached by the injectant, is directly related to the bulk transport of fuel in the air, and therefore it is an important precursor in the mixing processes. In a first approximation (Billig et al., 1970), the normal momentum exchange at the jet-injection station is the strongest among the physical interactions and heat transfer; shear or mixing can be neglected until the jet has gone through the Mach disk and has begun to turn in the axial direction. This approximation leads to an analysis based on conservation of average air and injectant properties, which can include the assumption of either jet Newtonian drag or isentropic jet momentum to obtain a correlation for the Mach disk height above the injection wall. This correlation, developed by Billig et al. (1970), emphasizes that penetration based on the Mach disk distance from the wall scales with the square root of the dynamic pressure ratio:

$$\frac{z}{D} \sim \left(\frac{q_j}{q_a}\right)^{0.5}, \tag{6.19}$$

where z is the distance above the wall, D is the jet-orifice diameter, and q_j and q_a are the dynamic pressures of the jet and the supersonic airstream, respectively. The penetration dependence on the dynamic pressure ratio was also maintained in models that define the jet penetration as the boundary of the region that contains injectant within a selected molar concentration, usually 99%–99.5% (Povinelli and Povinelli, 1971; Rogers, 1971; McClinton, 1974).

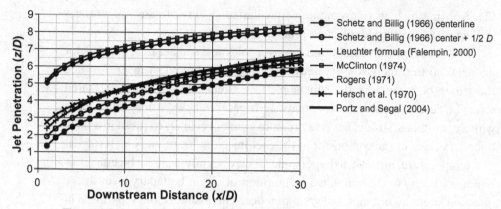

Figure 6.6. Jet penetration as a function of the axial location. These correlations indicate increased penetration as the air Mach number increases.

The power coefficient in approximation (6.19) that describes how the penetration depends on the dynamic pressure ratio varies from study to study, with values ranging from 0.3 to 0.56, and has the greatest effect on the jet penetration among the physical features usually selected in the penetration correlations; these parameters include the boundary-layer thickness at the injection location, the injectant molecular weight, and Mach number. In some cases, certain parameters are left out of the analysis, and the penetration correlation is simply related to the dynamic pressure ratio. Figure 6.6 includes a set of jet-penetration correlations from a number of studies, showing that variations exist between data sets obtained in different studies. Some researchers found twice the penetration height that others have measured. All of the studies, however, indicate a rapid penetration in the first 4–6 jet diameters followed by a slower penetration slope when the jet turns in a more axial direction relative to the airflow and the shear-layer development begins to dominate the mixing process. The empirical formulas derived in the studies and compared in Fig. 6.6 were based on the outer limit of the jet penetration, with the exception of the curves from Schetz and Billig (1966), which represent an analytically predicted transverse penetration at the centerline of an injected flow. In this model the injected plug of fluid carries transverse momentum into the airstream that initially has zero transverse momentum. This fluid particle is accelerated by momentum transfer from the airstream to the particle, and the airstream reacts by receiving and dissipating transverse momentum from the jet and converting some of its linear momentum to transverse momentum, primarily in the form of the spilled vortices.

The correlations based on Leuchter's formula (Falempin, 2000) and those of Hersch et al. (1970) and Portz and Segal (2004) agree well; however, there are differences between these results and those of other studies. The curve of Schetz and Billig (1966), which identifies the injected jet centerline whereas all

Table 6.1. *Constants in the penetration correlation in Eq. (6.20)*

Reference	A	B	C	E	F	G	Air Mach no.
Schetz and Billig (1966)	1	0.435	0	0.435	0	0	N/A
Leuchter (Falempin, 2000)	1.45	0.5	0.5	0.35	0	0	1.5
McClinton (1974)	4.2	0.3	0	0.143	0.057	0	4
Rogers (1971)	3.87	0.3	0	0.143	0	0	4
Hersch et al. (1970)	1.92	0.35	0.5	0.277	0	0	2
Portz and Segal (2004)	1.36	0.568	−1.5	0.276	0.2221	−0.0251	1.6

others identify the outer limit of gas penetration, is complemented by another curve marking $1/2$ jet diameter from the analytical centerline. This increment is maintained constant at all x/D for simplicity, as in the study by Billig et al. (1970). This modified curve of the Schetz and Billig (1966) analytical result approaches the correlations by Leuchter (Falempin, 2000), Hersch et al. (1970), and Portz and Segal (2004) in the far field, beyond $x/D = 20$.

One factor that influences the penetration measurement is the definition of the outer limit of penetration, which varies among studies. Near the penetration limit the concentration gradient is shallow, as shown by Rogers (1971); therefore any variation in definition can affect the measured penetration. However, this effect is small when the results of different studies are compared.

The test conditions used by each researcher mentioned in Fig. 6.6 affect the correlations they derived. For example, Leuchter's test condition of Mach 1.5 (Falempin, 2000) corresponded closely with the Mach 1.6 flows from which the correlation by Portz and Segal (2004) was derived; studies performed at Mach 2 free-stream conditions by Hersch et al. (1970) resulted in a slightly deeper prediction of penetration; and the studies at Mach 4 by Rogers (1971) and McClinton (1974) indicated further penetration increases.

A generic jet-penetration equation can be written as

$$z/D = A(q_j/q_a)^B(x/D + C)^E(\delta/D)^F(M_j/M_a)^G, \qquad (6.20)$$

where z is the penetration height, x is the axial distance from the jet-injection centerline, and M_j/M_a are the molecular weights of the jet and air, respectively. The constant and the exponents a, b, c, d derived in the studies quoted in Fig. 6.6 are shown in Table 6.1.

Equation (6.20) incorporates, in addition to the dynamic pressure ratio, the effects exerted on jet penetration by the boundary-layer thickness and the jet-to-air molecular-weight ratio. The wall boundary-layer thickness locally increases the dynamic pressure ratio, thus resulting in a larger penetration of the jet in the near field. As the jet becomes more axial, other mixing mechanisms, i.e., shear-layer development, dominate, and therefore the initial

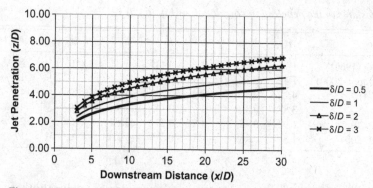

Figure 6.7. Effect of boundary-layer thickness at the jet-injection location. The jet penetration increases approximately proportional to the boundary-layer thickness in the near field. The far-field penetration is not affected by boundary-layer thickness. (Mach = 1.6, $q_j/q_a = 2$, $M_j/M_a = 0.070$.)

boundary-layer-thickness effect is less noticeable in the far field. The effects of boundary-layer thickness were included in the correlations by McClinton (1974) and Portz and Segal (2004). Figure 6.7 shows the boundary-layer effect found by Portz and Segal (2004) with measured boundary-layer thicknesses at the injection location in the range $\delta/d = 0.5$–3. The results indicated that the jet penetration increases approximately proportionally to the boundary-layer thickness in the near field but with no evident difference in the far field. This study found a stronger effect than did the study by McClinton (1974). The jet-to-air molecular-weight effect is an order of magnitude smaller, as shown in Fig. 6.8.

Other parameters were not found to play a significant role in the jet penetration. Povinelli and Povinelli (1971) found that the effect of the jet exit Mach number was relatively small compared with the air Mach number effect. The shape of the orifice itself does not offer substantial gains (Orth et al., 1969; Liscinsky et al., 1995) because the momentum exchange with the supersonic

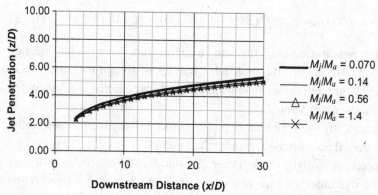

Figure 6.8. Molecular-weight ratio effect is an order of magnitude smaller than the other measured effects. (Mach = 1.6, $q_j/q_a = 2$, $\delta/D = 1$.)

airflow changes the initial jet shape rapidly following injection. Portz and Segal (2004) further found a negligible jet-to-air Re number effect. Therefore enhanced penetration for practical devices must rely on other means; some of these are discussed further in Subsection 6.3.4.

Because the penetration observed in the studies included in Fig. 6.6 increased as the air Mach number increased, it was suggested (Portz and Segal, 2004) to include this parameter in penetration Eq. (6.20) with the constants changed into functions of the Mach number. From the available databases Portz and Segal (2004) suggested the following relations:

$$A = 1.05\,M_{air} - 0.192,$$
$$B = -0.0803\,M_{air} + 0.615,$$
$$C = -2.34\,/M_{air},$$
$$E = 0.406\,M_{air}{}^{(-0.823)},$$
$$F = -0.067\,M_{air} + 0.325,$$
$$G = -0.0251.$$

These correlations satisfy all the test results compiled in the studies quoted in Fig. 6.6 in which the inputs to the formula, including Mach number, boundary-layer thickness, and dynamic pressure ratio, are taken directly from test data compiled in these studies. Figure 6.9 shows that the prediction based on Eq. (6.20) satisfies the other correlations. The figure also includes a comparison with test data obtained for $M_{air} = 2.5$, which was not used to create the formula.

Generally, good agreement is evident, particularly in the midfield. Close to the injection location where the slopes are large or in the far field, beyond 30 jet diameters, the differences are 1 to 2 jet diameters.

An examination of Eq. (6.20) modified with Mach-dependent exponents shows that penetration increases with increased dynamic pressure ratio, downstream distance, and boundary-layer-thickness-to-jet-diameter ratio, as expected. The penetration also increases as the air Mach number increases, which is not intuitively expected. The increased penetration that is due to larger Mach numbers can be attributed to the stronger bow shock structure associated with higher Mach flows. The dynamic pressure on the downstream side of the shock is an increasingly small percentage of the upstream dynamic pressure as the Mach number increases, so the dynamic pressure ratio actually "experienced" by the injected gas is much lower than the free-stream value. For example, a Mach 4 airstream passing through a normal shock will experience a dynamic pressure reduction of 78%. However, direct application of a correction based on the normal shock is not sufficiently accurate to correct the dynamic pressure ratio effect because the bow shock is normal over only a portion of the plume.

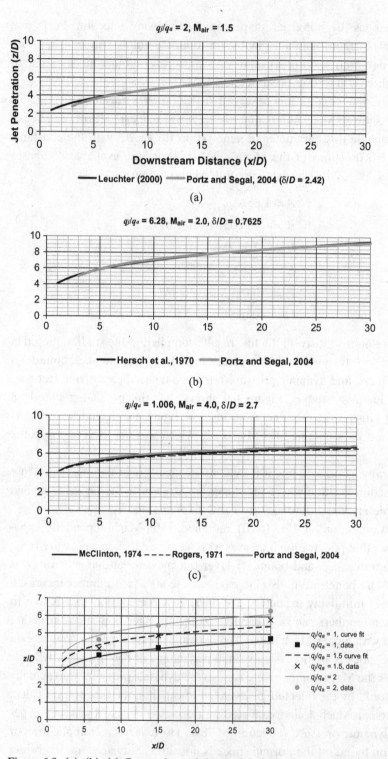

Figure 6.9. (a), (b), (c) Comparison of the modified penetration formula with existing correlations; (d) comparison with experimental data not included in the derivation of the modified formula.

Despite increased penetration from the wall when the Mach number is low and the boundary layer is thick, the penetration out of the boundary layer and into the free stream is decreased by a thick boundary layer. The strong, nearly normal bow shock is absent. The supersonic free stream interacts with the deflected plume by forming a weaker oblique shock. With the weakened virtual obstruction of the injectant jet in the free stream, the vortex generation essential to effective bulk mixing processes is weakened. The resulting flow approaches the case of tangential, rather than transverse, injection.

When the supersonic Mach number is high, the bow shock is strong, with a consequent low dynamic pressure downstream of the bow shock and increased penetration. This explains the decreased effect of boundary-layer thickness on improving penetration as the Mach number increases.

TRANSVERSE LIQUID JETS. The issues of jet penetration and mixing apply as well when liquid fuels instead of gases are used in the scramjet engine. There are clear technological, economical, and operational advantages in requiring liquid-fuel use in comparison with gaseous-based systems, particularly when liquid-hydrocarbon formulations are applied for small hypersonic vehicles limited to Mach 8 flight. Some of these advantages were discussed in Chap. 4. However, the multistage physical–chemical processes of liquid-hydrocarbon-fuel mixing and burning increase the requirements for fast mixing when the short residence time in the scramjet engine is considered. If the selected fuel is amenable to operating at supercritical conditions, a significant gain in the time required for the liquid-fuel breakup and vaporization can be achieved; furthermore, if chemical decomposition accompanies these transformations, the potential formation of hydrogen or other active radicals will result in increased reactivity of the mixture and a reduction of the combustion length.

The interaction between the liquid jet and the supersonic airstream is dominated by the instabilities that develop on the surface of the liquid column, resulting in jet breakup, vaporization, and mixing (Schetz, 1980; Chigier and Reitz, 1996). These instabilities appear as a result of aerodynamic forces and develop and increase along the jet trajectory until the jet breaks into irregularly shaped clumps of liquid material. Kush and Schetz (1973) identified distinct regimes of liquid jet penetration and breakup, depending on the dynamic pressure ratio. At high dynamic pressure ratios, with values in excess of six, the liquid jet penetrates for several jet diameters undisturbed, then begins turning downstream in the supersonic flow direction, and forms a regulate pattern of surface waves that grow in amplitude, finally leading to jet breakup. At intermediate dynamic pressure ratios, between 1.5 and 6, large jet surface waves develop immediately on injection with an unsteady shape or frequency, and the jet disintegrates within several jet diameters. Below these dynamic pressure ratios, the jet is abruptly turned by the high-momentum airflow and remains in a narrow layer close to the injection wall, exhibiting random spatial and temporal shape disintegration. The breakup regime is dominated by aerodynamic forces in the initial stage but, as the waviness of the jet surface increases and large structures are formed, the breakup is dominated by liquid turbulence and inertial forces, ultimately leading to jet disintegration (Fuller et al., 2000).

Including the jet shape and the angle of the liquid injection, Schetz (1980) offered the following liquid jet-penetration correlation:

$$\frac{P}{D} = \text{const} \left(\frac{q_j}{q_a}\right)^{0.5} (\text{AR})^{0.46} \ln\left[1 + 6\left(\frac{x}{D}\right)\right] \sin\left(\frac{2\theta}{3}\right), \qquad (6.21)$$

where AR is the jet-orifice aspect ratio and θ is the injection angle. It should be noted that, unlike that of the gaseous injection, the shape of the liquid jet plays a role in the liquid jet-penetration process because it is closely related to the formation of the surface disturbances that lead to jet disintegration. Aside from the influence that the instability regimes previously described have on the liquid jet surface, the jet penetration scales with the square root of the dynamic pressure ratio, similar to the gaseous jet case.

6.3.3 Degree of Mixing and Mixing Efficiency

The jet penetration is a global measure that promotes fuel–air mixing. In fact, the jet-mixture fraction distribution is of more interest because it helps identify the regions where sufficient mixing has occurred to enable the initiation and propagation of chemical reactions. In a simple, two-component system, the concentration is defined by the mass-flow ratio. A mixedness parameter was defined by Liscinsky et al. (1995) as

$$U = \frac{c_{\text{var}}}{c_{\text{avg}}\left(1 - c_{\text{avg}}\right)}, \qquad (6.22)$$

where c_{var} is a spatial concentration variance defined as

$$c_{\text{var}} = \frac{1}{n} \sum_{i=1}^{n} \left(\bar{c}_i - c_{\text{avg}}\right)^2,$$

$\bar{c}_i$ is the time-averaged concentration at any given location, and the average concentration c_{avg} determined from the global mass flows of the jet and the airstream as $c_{\text{avg}} = [m_{\text{jet}}/m_{\text{jet}} + m_{\text{air}}]$.

This normalization by the product $c_{\text{avg}}(1 - c_{\text{avg}})$ removes the dependence on the jet-to-air mass-flow ratio. As defined by approximation (6.19), the mixedness parameter U varies from 0 for a perfectly mixed system to 1 for completely segregated components. The local concentration $\bar{c}_i$ can be obtained either from computation or from experiment. It would appear more intuitive to use $1 - U$ as a parameter, as chosen by Fuller et al. (1998), because

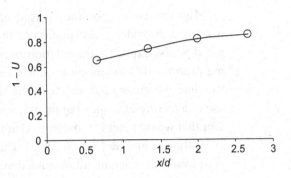

Figure 6.10. Degree of mixedness of a gaseous jet in a subsonic airstream with $q_j/q_a = 8.2$ (Liscinsky et al., 1995).

it describes increasing mixedness. Shown in Fig. 6.10 is $1 - U$ for the concentration measurements performed by Liscinsky et al. (1995) for a transverse jet in a subsonic airstream with dynamic pressure ratio $q_j/q_a = 8.2$. The degree of mixedness increases in the downstream direction as expected; the degree of mixedness dependence on the axial location clearly depends on other parameters, which are not included in the figure. Most of all, the injection configuration affects this parameter and, as a result, the mixedness calculated by Fuller et al. (1998) for angled, supersonic jet injection from a ramp in a Mach 2 airstream differs substantially from the results obtained by Liscinsky et al. (1995); the exponent in the power correlation in the study by Fuller et al. (1998) is approximately one order of magnitude lower than for the correlation shown in Fig. 6.10. The more modest mixing development in the study of Fuller et al. is clearly due to the injection configuration that used an angled jet exit approaching the shear-layer-dominated mixing, which was seen to delay mixing with increased compressibility. The particular case of ramp injection has attracted substantial attention because it offers advantages over either the parallel or the transverse injection solutions. This configuration is discussed in more detail in a following section.

The concentration variation described by the mixedness parameter U indicates the degree to which the two components in the system are present within a given volume in the flow. In a sense, this parameter is not unlike the *nonuniformity mass-fraction* parameter introduced by Kopchenov and Lomkov (1992), defined as

$$D = \frac{\int_A \rho u (c - \bar{c})^2 \, dA}{\bar{c}^2 \int_A \rho u \, dA},$$

(6.23)

where D is the nonuniformity mass fraction; ρ, u, and c are the local density, velocity, and concentration, respectively; A is the cross section of the axial station where mixing is evaluated; and $\bar{c}$ is the mass-averaged concentration in the cross section. A value of $D = 0$ indicates full uniformity and $D = 1$ indicates complete lack of injectant. The first parameter, U, is a local concentration measurement whereas the second, D, offers a cross-sectional measure.

The degree of mixedness based on concentration decay, as previously described, provides a distribution of the injectant in the airstream but does not describe explicitly the relation between the concentration and the mixing ratio needed to sustain chemical reactions. For this purpose it is useful to relate the mixing parameter to the stoichiometric ratio. This suggests the use of a *mixing-efficiency* parameter, which indicates the *fraction* of the reactant that would react if brought to chemical equilibrium with the air (Rogers, 1971; Riggins and McClinton, 1992). The fraction of the reactant refers to the least-available reactant, air, or fuel, depending on whether the mixture is lean or rich; in fuel-lean regions, the mixing-efficiency parameter represents the fraction of fuel, and in fuel-rich regions the mixing efficiency refers to the fraction of air. The fuel fraction defined in this fashion takes the following values,

$$\alpha_{react} = \begin{cases} \alpha & \text{for } \alpha \leq \alpha_{stoic} \\ \alpha(1-\alpha)/(1-\alpha_{stoic}) & \text{for } \alpha > \alpha_{stoic} \end{cases}, \tag{6.24}$$

where α is the fuel mass fraction, α_{react} is the fuel fraction mixed in a proportion that can react, and α_{stoic} is the fuel stoichiometric mass fraction. The mixing efficiency is then defined as

$$\eta_m \equiv \frac{\dot{m}_{fuel,mixed}}{\dot{m}_{fuel,total}} = \frac{\int \alpha_{react} \rho u \, dA}{\int \alpha \rho u \, dA}, \tag{6.25}$$

where $\dot{m}_{fuel,mixed}$ is the mixed fuel mass flow and $\dot{m}_{fuel,total}$ is the total fuel flow rate. Equation (6.25) thus defines a mixing efficiency in a cross section, with $\eta_m = 1$ indicating a perfectly mixed system. In this case the maximum value of fuel fraction must remain less than or equal to the stoichiometric ratio.

The mixing efficiency changes with an axial direction, depending on the geometric configuration and the air–fuel system thermodynamic properties. Several correlations have been suggested (Rogers, 1971; Fuller et al., 1998) by formulation of empirical equations from measured concentrations. Common to these correlations is the power dependence on the axial distance from the injector and, in some cases (Rogers, 1971), the dependence on the dynamic pressure ratio. Many other factors, including the number of injectors and their spacing, the fuel molecular weight, the injection and combustion chamber configuration, the presence of shock waves, etc., play a role in determining the mixing efficiency in a supersonic combustion flow. As a result, the numerical coefficients that appear in the equations formulated in the studies previously mentioned vary according to the injection parameters.

A particular working formula was offered in the early studies at NASA Langley Research Center (Northam and Anderson, 1986) for transverse injection from two round orifices on opposite combustor walls, placed such that the spacing between the injectors on each side equals half the duct width. In this formulation it is assumed that the mixing is complete for a stoichiometric fuel–air mixture at a distance of $X = 60G$, where G is the height of the duct. For mixtures at different ratios than stoichiometric, the following empirical equation is offered for the complete mixing length:

$$X_\phi = \begin{cases} 3.333 \ \exp(-1.204\phi) & \text{for } \phi > 1 \\ 0.179 \ \exp(1.72\phi) & \text{for } \phi \le 1 \end{cases}. \tag{6.26}$$

The mixing efficiency is then related to the axial distance from the injector's center as a function of the mixing length:

$$\eta_m = 1.01 + 0.176 \ln\left(\frac{X}{X_\phi}\right). \tag{6.27}$$

If the wall injection results in large momentum losses and the axial injection requires long distances to achieve the molecular mixing required to initiate chemical reactions, other means are needed to accelerate mixing. A few of the possible solutions are subsequently described.

6.3.4 Mixing Enhancement

A simple evaluation of a gaseous fuel-ignition time t_i shows that, under typical thermodynamic conditions met in a supersonic ramjet combustor, i.e., pressure $p = 50$–100 kPa and temperature $T = 600$–1000 K, $t_i = 5$–10 ms (Goltsev et al., 1991). This is a long time compared with the residence time estimated to only a few milliseconds. Therefore simple fuel-injection methods are not sufficient to ensure that all processes involved in heat deposition in the combustion chamber would take place to completion. Furthermore, additional processes appear if liquid fuel is used, including jet breakup and vaporization, hence compounding the residence time deficiency. The addition of active chemical components to the fuel, such as hydrogen or ClF_3, SiH_4, or other compounds as chemical-reactions promoter, can reduce the induction time (Diskin and Northam, 1986; Hunt et al., 1982; Waltrup, 1987). This method is effective for accelerating the combustion process initiation, but it does not provide a feasible solution for the entire flight duration, as the proportion of added components may be as high as 25% of the entire mission fuel requirement (Bonghi et al., 1995), thus increasing the vehicle volume and the complexity and, in some cases, causing a certain decrease in the specific impulse. It appears therefore that chemical-reaction acceleration options are limited, and significant gains must be achieved through mixing-enhancement measures. These measures include selection of fuel-injection configurations and the generation of jet–air fluid mechanics interactions.

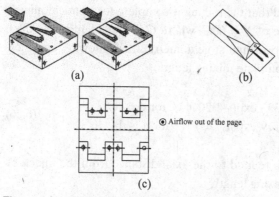

Figure 6.11. Ramp-injection configurations. The airflow spills over the ramps, forming vortical structures that lift the fuel from the wall and enhance mixing. (a) Swept and straight ramps penetrating the flow field (Northam et al., 1992). The swept ramps are more effective than the unswept ramps, with a calculated combustion efficiency approaching that of transverse sonic injection. (b) Recessed swept ramp occupying the entire duct with multiple injection ports (Owens et al., 1997). (c) Multiple ramps with a partially recessed wall to maintain a constant-cross-section area with multiple injection ports (Goldfeld et al., 2004).

RAMPS. To avoid the strong shocks associated with transverse injection from the wall, a low angle of fuel injection is desirable. The difficulty with this selection is that the fuel remains close to the wall, and therefore much of the core airflow cannot participate in the mixing process until far downstream, where the shear layers have sufficiently developed. An intermediate solution is therefore expected to provide enhanced mixing without incurring unacceptable levels of pressure losses (Swithenbank et al., 1989).

Injecting the fuel from ramps facing downstream represents such a solution, and some of the suggested ramp configurations (Northam et al., 1992; Hartfield et al., 1994; Riggins et al., 1995; Owens et al., 1997) have shown reasonable levels of far-field mixing when compared with transverse injection alternatives. Figure 6.11 includes schematics of several types of possible ramp configurations, including protruding ramps with swept or unswept sides (Northam et al., 1992), recessed ramps with multiple injection ports (Owens et al., 1997), or complex configurations, such as the one studied by Goldfeld et al. (2004), which includes partially protruding ramps and partially recessed walls to maintain the duct constant-cross-section area and staggered layout on opposite chamber walls. Small recirculation regions form between the ramp's downstream face and the combustion chamber wall and help promoting flame-holding.

The major mixing-enhancing mechanisms produced by a simple straight or swept ramp, such as those shown in Fig. 6.11(a), are the pairs of axial counter-rotating vortices that form as the air flows over the ramps. The ramp vortex shedding helps lift the fuel from a low injection angle and promotes penetration

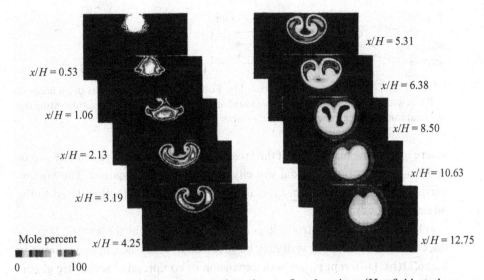

Figure 6.12. Injectant mole fraction at selected cross-flow locations (Hartfield et al., 1994). At two ramp heights downstream of the ramp face, the vortices' effects on the fuel jet become visible, with fuel material entrained by these large structures; the fuel plume begins to be lifted from the wall. Beyond eight ramp heights, the plume concentration is everywhere below the stoichiometric ratio.

into the core airstream. The formation of the counterrotating vortices and their interaction with the fuel jet are visible in the cross-flow mole-fraction images obtained by Hartfield et al. (1994), as shown in Fig. 6.12. The ramp used in this study is of the straight type shown in Fig. 6.11(a). Close to the ramp face, the jet maintains the round shape of the injection orifice but within a short distance, slightly beyond two ramp heights, the ramp-induced vortices' effects on the fuel jet become visible with fuel material entrained by these large structures, and the fuel plume begins to be lifted from the wall; beyond eight ramp heights, the plume concentration is everywhere below the stoichiometric ratio.

The swept ramps shown in Fig. 6.11(a) have proven more effective to enhance mixing through generation of stronger vortices than the unswept ramps (Northam et al., 1992; Riggins and Vitt, 1995). Calculated combustion efficiencies achieved with the swept ramp injectors were higher in comparison with those achieved with the unswept ramp injectors and approached the calculated values for the perpendicular sonic injection mixing model (Northam et al., 1992).

The vortex generated by the swept ramps increases the lateral and vertical stretching of the fuel–air interface and may increase the mixing efficiency by as much as 25% more than the unswept ramp (Riggins and Vitt, 1995). The different vortex strengths in the two cases led to the concept of the *vortex-stirring length* suggested by Riggins and Vitt (1995), defined as

$$L_{vs} = \int \frac{|\bar{q}_{cross}|}{U_{avg}} dx, \tag{6.28}$$

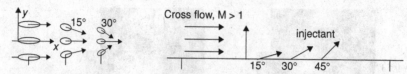

Figure 6.13. The aerodynamic ramp used by Fuller et al. (1998) consists of an array of wall jets with various injection angles in both axial and lateral directions, simulating the physical ramp vortex generation by a number of fuel sources.

where q_{cross} is the magnitude of the cross-flow component of the velocity vector and U_{avg} is the effective axial velocity at a given axial location. The vortex-stirring length, which is larger for the unswept ramp, is directly related to the mixing efficiency.

The ramps may protrude into the flow, as shown in the Fig. 6.11(a), or the combustion chamber wall may be recessed to form the ramp, as shown in Fig. 6.11(b). Different patterns of expansion or compression waves are generated in each case, and these waves interact with the mixing layers as the flow propagates in an axial direction. The misaligned density and pressure gradients create vorticity, as evidenced by the vorticity transport equation:

$$\frac{d\bar{\omega}}{dt} = \nabla p \times \nabla\left(\frac{1}{\rho}\right).$$

(6.29)

In turn, the increased vorticity further contributes to enhance mixing (Marble et al., 1990; Drummond, 1997).

A notable approach to the generation of vortical structures is the *aerodynamic ramp*, such as the one suggested by Fuller et al. (1998) and shown in Fig. 6.13. This concept consists of an array of wall injectors with various injection angles in both axial and lateral directions, simulating the physical ramp vortex generation by a number of fuel sources. Mixing is enhanced by the multiple vortex–injectant interactions. The aerodynamic ramp eliminates the cooling requirements of physical ramps, especially in localized hot spots such as in recirculation regions. At the same time, the aerodynamic ramp is expected to reduce the drag while maintaining similar far-field mixing characteristics (Fuller et al., 1998).

UPSTREAM FUEL INJECTION – INCREASING THE RESIDENCE TIME. To increase the fuel residence time, thus achieving a higher degree of mixing in the combustion chamber, it may be useful to inject part of the fuel upstream of the combustion chamber in the isolator or even in the inlet or further upstream on the vehicle forebody. Some of the concepts of forebody fuel injection were suggested in conjunction with the generation of strong shocks at the inlet capture to operate in a detonation-wave ramjet mode (Sislian, 2000). Clearly this method introduces additional interactions between the fuel injection and the hypersonic layer formed on the vehicle forebody, with important implications for the vehicle stability, in particular under off-design conditions, and may modify the flow around the inlet capture.

Fuel injection in the inlet is less problematic than injection from the forebody in that sense and has shown the potential to decrease the mixing length by as much as a half for the typical ramjet with a design Mach number of around 6 (Vinogradov and Prudnikov, 1993). Furthermore, other mixing-enhancement mechanisms are present in the inlet such as shock–jet interactions (Vasiliev et al., 1994). The fuel mixing with the inlet airflow, and the additional processes that may be present if liquid fuels are used, i.e., jet breakup and vaporization, may lead both to degradation or improvement of the inlet starting and operational changes, depending on the fuel type and rate, and the flight conditions.

Several practical issues arise in the case of preinjection of liquid fuel in the inlet duct including (a) balancing mixing efficiency, flow deceleration, and inlet performance, and (b) the ability to avoid flashback by eliminating the residence of the fuel on the inlet–isolator walls. The fuel can be injected from struts or pylons that protrude into the flow and facilitate fuel distribution far from the walls and with good spatial uniformity into the core airflow (Gruenig et al., 2000; Sabel'nikov and Prezin, 2000; Parent and Sislian, 2003). These pylons are intrusive, occupy a nonnegligible area, and, in addition to the pressure losses they generate, they require cooling. A particular type of pylon, which reduces or even eliminates the deficiencies that accompany the struts' presence in the flow, is a thin pylon with the fuel injected from the wall at a short distance behind it. These pylons create a region of low pressure behind them that has the effect of lifting the fuel from the walls, thus avoiding the penetration of a combustible mixture in the boundary layers. The problem of flashback can thus be eliminated. A more detailed discussion of these pylons is included in Chap. 5 based on the review by Vinogradov et al. (2007).

The combustion studies (Vinogradov et al., 1990; Guoskov et al., 2001) have shown that the injection of hydrogen behind the pylons at different experimental conditions did not result in premature ignition and upstream interaction. The same is true for liquid-fuel injection, as seen in Fig. 6.14, taken from the experiments by Livingston et al. (2000). In an inlet operating at $M = 3.6$, the pylon shown in Fig. 6.14(a) helps to remove the fuel jet entirely away from the wall. The jet experiences an abrupt breakup and is carried into the inlet core airflow at the pylon height. The cross-flow light-scattering image in Fig. 6.14(b) shows that the pylons (two pylons side by side have been used in these experiments) is effective in keeping the fuel in the core of the inlet's flow, whereas in the absence of the pylon, the jet spreads rapidly to the inlet wall, as shown in Fig. 6.14(c).

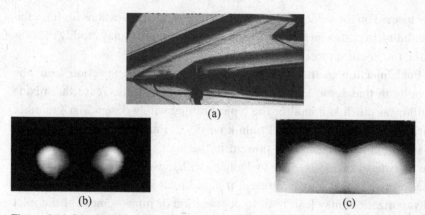

Figure 6.14. Liquid injection behind a thin pylon (Livingston et al., 2000): (a) the pylon creates a low-pressure region that enables the jet to be lifted above the wall and to penetrate the flow to the pylon height; (b) cross-flow light scattering with pylons shows the fuel concentrated in the airflow core; and (c) in the absence of the pylon, the jet reaches the inlet duct walls.

There are other parameters characteristic of the inlet flow that, along with the fuel-injection scheme and the duct geometry, help to prevent flashback: The temperature and pressure are low, and therefore the ignition delay time is long. For example, the fuel residence time at the static flow parameters typical for an operating at $M_{flight} = 3.6$ is 0.4 ms, far shorter than the expected ignition delay time for hydrocarbon fuels at this temperature and pressure. It is thus possible to achieve a certain degree of mixing without ignition within the inlet.

From an operational point of view, it has been noted that certain beneficial effects could be obtained from the presence of fuel along the walls such as film cooling. These beneficial effects are overcome by the potential danger of flashback, and, in this case, the presence of fuel in the boundary layers along the walls of the inlet and isolator is undesirable.

AERATED INJECTORS. Liquid fuels must undergo additional physical processes before molecular-level mixing in comparison with gaseous fuels, including jet breakup and vaporization. These processes can be accelerated by injection of a gas concentric with the liquid jet to generate an effervescent liquid at the injection station, as suggested by Sabel'nikov and Prezin (2000) in the schematic shown in Fig. 6.15. The gas can be either air or a gaseous fuel, for example, hydrogen. The method accelerates the jet breakup and, along with it, turbulent mixing, leading to a shorter combustion length. However, along with the additional fuel system complexities, gas addition to the liquid fuel reduces the fuel energy density. Clearly, if the gas used is itself a fuel, it will participate in heat generation in the combustion chamber but under certain conditions air injection may prove beneficial by reducing the local equivalence ratio and promoting combustion and flameholding.

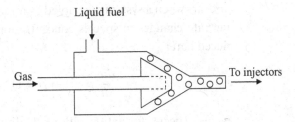

Figure 6.15. Aerated liquid-fuel injector with a concentric gas (after Sabel'nikov and Prezin, 2000).

6.4 Chemical Kinetics – Reaction Mechanisms

In a system in which the reaction times are long compared with the residence time, the complete chemical transformation must be treated as a finite-rate process with a series of steps, each involving an elementary chemical reaction. Knowing the specific reaction rates of these elementary steps allows building a complete mechanism that describes the fuel–oxidant reaction. Detailed reaction mechanisms for any system, except a limited number of very simple compounds, may include a computationally unmanageable number of elementary reactions. For example, the Gas Research Institute (GRI) mechanism (Smith et al., 1999) includes hundreds of reactions, even for simple hydrocarbon systems. For larger hydrocarbon systems, thousands of reactions are needed. Moreover, for systems including larger hydrocarbon systems, such as heptane and beyond, only reduced mechanisms have been developed so far (Li et al., 2001), and detailed reactions mechanisms do not exist. A proposed reaction mechanism is validated by matching the species production–decay to experimental results obtained in controlled environments (such as ignition delay time measurements in shock tubes; Colket and Spadaccini, 2001; Davidson et al., 2001; Mikolaitis et al., 2003). The selection of the reaction mechanisms is based mostly on past experience and limited selection rules that prohibit certain reactions from taking place. Mechanisms reduced to only tens of reactions for hydrocarbon systems or to only several reactions for the simpler systems, such as hydrogen–air chemistry, are based on adapting the reaction rates to match the results of the more detailed reaction mechanisms. Finally, reduced reaction mechanisms usually correspond to the domain of pressure, temperature, and concentrations for which they have been established; once a reaction mechanism is validated against existing data, selection of a subset of elementary reactions from a reaction mechanism for simplified calculations without adjusting the reaction rates for the new reduced reaction mechanism may lead to nonnegligible errors.

Several reaction mechanisms for hydrogen–air and simple hydrocarbon systems that have been applied for supersonic combustion studies are subsequently described. Although the expansion of computational capabilities would suggest that detailed reaction mechanisms can increasingly be used, solving a complicated system of equations may be difficult. The problem appears when a system of coupled differential equations, such as those describing rate changes in species concentration – described in Chap. 2 and reproduced here:

$$\frac{dn_i}{dt}/(v_i'' - v_i') = \omega_i, \quad i = 1, \ldots, N, \tag{6.30}$$

include species formation with vastly different time scales, resulting in what is termed a *stiff* system of differential equations (Oran and Boris, 2001; Poinsot and Veynante, 2001). Reduced reaction mechanisms are then used either by removing certain reactions that progress much faster than the others, using the partial-equilibrium method, or by assuming that certain species are in steady state. Certain reactions can thus be removed from the detailed mechanism, generating a reduced reaction mechanism.

The steady-state assumption, which is subsequently invoked to describe an example of a reduced reaction mechanism for the hydrogen–oxygen system, is based on the observation that the diffusion time for a particular chemical species is long compared with the chemical time associated with that species (Williams, 1985). In this case the net rate change of the species assumed in equilibrium, that is, the difference between the rate of species formation and destruction, is negligible compared with the rate of formation, i.e, $(\omega_{i+} - \omega_{i-})/\omega_{i+} \ll 1$.

The partial-equilibrium assumption refers to particular reactions characterized by equal forward and reverse reaction rates. This assumption leads to the elimination of a differential equation from the system and replacing it with an algebraic equation. It is a particularly useful method when disparate chemical times exist between the reactions in the mechanism, leading to what is called *stiffness* of the differential equation system. The steady-state assumption is based on identification of the species at steady state for the given chemical system and the thermodynamic conditions in the region of analysis.

6.4.1 Hydrogen–Air Reaction Mechanisms

Because hydrogen is an important fuel candidate for scramjet applications, in particular for the high-Mach-number flight regime, its chemistry has attracted substantial attention. The relatively small number of species involved in comparison with hydrocarbon systems made these studies also more amenable both from the point of view of experimental determination of reaction rates and from the analysis of detailed reaction mechanisms. Studies such as those of Correa and Mani (1989), Harradine et al. (1990), and Sangiovanni et al. (1993) are only a few examples that attempted to identify chemical-reaction mechanisms specifically suitable for the environment found in a supersonic combustion chamber or even in a scramjet's nozzle where chemical reactions may still take place because of the short residence time and slow recombination rates. Detailed mechanisms involving 18–21 reactions in hydrogen–oxygen chemistry have been proposed (for example, see the 19 reaction mechanisms of Yetter et al., 1991) and extended to more than 50 reactions when nitrogen chemistry is also considered. Often cited for the hydrogen–air-system scramjet computation is Jachimowski's (1988) 33-reaction mechanism, which is listed in Table 6.2. This reaction mechanism has been verified for a broad range of thermodynamic conditions expected in a scramjet engine, from Mach 8 to 25.

Reaction 1, the decomposition of molecular hydrogen and oxygen to OH, is the chain-initiating reaction because it forms the free radicals that contribute to the initiation and propagation of the other reactions. The activation energy is quite high compared with other reactions in the mechanism. The energy required for dissociating the H_2 and O_2 molecules is provided through energetic collisions. The OH radicals contribute through reaction 4 to the formation of atomic hydrogen that in turn contributes to formation of atomic oxygen in chain-branching reaction 2. Atomic H and atomic O are continually generated in the mechanism and are therefore chain carriers in this mechanism. Reaction 19, the decomposition of H_2O_2 into 2OH, is also a chain propagation reaction, and it is sensitive in particular at higher pressures and temperatures. The formation of H_2O_2 follows from reaction 15, the decomposition of HO_2, an intermediate species that has a short residence time at high temperatures and contributes to the chain propagation through reactions 15 and 19, but it plays a chain-terminating role at low pressures and temperatures, close to the second explosion limit (Glassman, 1996), when it is long lived and may carry away energy and dissipate it to a wall; the process of HO_2 formation through reactions 9, 10, and 14 reduces the pool of free radicals, and it may be found in higher concentrations than H, O, or OH (Nishioka and Law, 1997) in low-temperature regions. Because these conditions exist in certain regions, the scramjet engine HO_2 must be accounted for even if a reduced reaction mechanism is used in the analysis; otherwise the solution may result in unrealistically high rates of heat generation (Segal et al., 1995). Formation of HO_2 through reaction 9 competes with chain-branching reaction 2 for ignition delay time, which may change by an order of magnitude for a mixture temperature change from 1000 to 975 K (Jachimowski, 1988).

The ignition study by Nishioka and Law (1997) makes particularly evident the competition between chain-branching reactions 2 and 3 and chain-terminating reaction 9. Figures 6.16(a) and 6.16(b) adopted from their study indicate the streamwise mole fraction and temperature distribution for a simple shear layer developing between initially separated air and hydrogen flows originating from a splitter plate at $x = 0$ in the figures. The difference in the initial temperature from 800 K in Fig. 6.16(a) to 1200 K in Fig. 6.16(b) clearly indicates the higher HO_2 concentration at the lower temperature and the

Table 6.2. *Jachimowski's (1988) hydrogen–air reaction mechanism*

	Reaction	A	n	E
1	$H_2+O_2 \rightarrow OH+OH$	1.70E + 13	0	48 000
2	$H+O_2 \rightarrow OH+O$	2.60E + 14	0	16 800
3	$O+H_2 \rightarrow OH+H$	1.80E + 10	1	8900
4	$OH+H_2 \rightarrow H_2O+H$	2.20E + 13	0	5150
5	$OH+OH \rightarrow H_2O+O$	6.30E + 12	0	1090
6	$H+OH+M \rightarrow H_2O+M$	2.20E + 22	-2	0
7	$H+H+M \rightarrow H_2+M$	6.40E + 17	-1	0
8	$H+O+M \rightarrow OH+M$	6.00E + 16	-0.6	0
9	$H+O_2+M \rightarrow HO_2+M$	2.10E + 15	0	-1000
10	$HO_2+H \rightarrow H_2+O_2$	1.30E + 13	0	0
11	$HO_2+H \rightarrow OH+OH$	1.40E + 14	0	1080
12	$HO_2+H \rightarrow H_2O+O$	1.00E + 13	0	1080
13	$HO_2+O \rightarrow O_2+OH$	1.50E + 13	0	950
14	$HO_2+OH \rightarrow H_2O+O_2$	8.00E + 12	0	0
15	$HO_2+HO_2 \rightarrow H_2O_2+O_2$	2.00E + 12	0	0
16	$H+H_2O_2 \rightarrow H_2+HO_2$	1.40E + 12	0	3600
17	$O+H_2O_2 \rightarrow OH+HO_2$	1.40E + 13	0	6400
18	$OH+H_2O_2 \rightarrow H_2O+HO_2$	6.10E + 12	0	1430
19	$M+H_2O_2 \rightarrow OH+OH+M$	1.20E + 17	0	45 500
20	$O+O+M \rightarrow O_2+M$	6.00E + 17	0	-1800
21	$N+N+M \rightarrow N_2+M$	2.80E + 17	-0.75	0
22	$N+O_2 \rightarrow NO+O$	6.40E + 09	1	6300
23	$N+NO \rightarrow N_2+O$	1.60E + 13	0	0
24	$N+OH \rightarrow NO+H$	6.30E + 11	0.5	0
25	$H+NO+M \rightarrow HNO+M$	5.40E + 15	0	-600
26	$H+HNO \rightarrow NO+H_2$	4.80E + 12	0	0
27	$O+HNO \rightarrow NO+OH$	5.00E + 11	0.5	0
28	$OH+HNO \rightarrow NO+H_2O$	3.60E + 13	0	0
29	$HO_2+HNO \rightarrow NO+H_2O_2$	2.00E + 12	0	0
30	$HO_2+NO \rightarrow NO_2+OH$	3.40E + 12	0	-260
31	$H+NO_2 \rightarrow NO+OH$	3.50E + 14	0	1500
32	$O+NO_2 \rightarrow NO+O_2$	1.00E + 13	0	600
33	$M+NO_2 \rightarrow NO+O+M$	1.16E + 16	0	66 000

Notes: The specific reaction-rate coefficients are given in Arhennius form, $k = AT^n \exp(-E/RT)$ with units of seconds, moles, cubic centimeters, calories, and degrees Kelvin. The rates for the reversed steps are obtained from NIST-JANNAF thermochemical equilibrium data. The third-body efficiencies are as follows:

reaction 6, N_2: 1, H_2O: 6.0;
reaction 7, N_2: 1, H_2: 2.0 and H_2O: 6.0;
reaction 8, N_2: 1, H_2O: 5.0;
reaction 9, N_2: 1, H_2: 2.0, and H_2O: 16.0;
reaction 19, N_2: 1, H_2O: 15.0.

tapering of HO_2 production at the higher temperature. Following ignition and a temperature rise, an equation of type 11 begins to play a role and HO_2 concentration decays. "High-temperature chemical kinetics" becomes dominant (Nishioka and Law, 1997).

Table 6.3. *Reactions with O_2 ($^1\Delta_g$) included in the hydrogen–oxygen mechanism by Starik and Titova (2001)*

	Reaction	A	n	E/R
30	O_2 ($^1\Delta_g$)+M→2O+M	2.60E + 18	0	−48 188
31	HO_2+M→O_2 ($^1\Delta_g$)+H+M	$q_a \times$ 2.1E + 15	0	−23 000
32	2HO_2→H_2O_2+O_2 ($^1\Delta_g$)	$q_a \times$ 1.8E + 13	0	−500
38	H+O_2 ($^1\Delta_g$)→OH+O	1.10E + 14	0	−3188
39	H_2+O_2 ($^1\Delta_g$)→2OH	1.70E + 15	0	−17 080
40	H_2+O_2 ($^1\Delta_g$)→H+HO_2	2.10E + 13	0	−18 216
41	H_2O+O_2 ($^1\Delta_g$)→OH+HO_2	1.50E + 15	0.5	−25 521
42	OH+O_2 ($^1\Delta_g$)→O+HO_2	1.30E + 13	0	−17 007
43	H_2O_2+O→H_2O+O_2 ($^1\Delta_g$)	$q_a \times$ 8.4E + 11	0	−2130
45	O_2 ($^1\Delta_g$)+M→O_2 +M	3.36E + 06	0	0

Reactions of types 6 and 14 are recombination reactions that form stable products and are chain-terminating reactions. The three-body reactions, such as 6, 7, and 20, require the presence of a third body to carry away the energy that otherwise would render the product unstable and able to dissociate.

Reactions 20–33 include nitrogen-containing species that may become important at higher temperatures, namely, at higher flight Mach numbers.

Interesting additions to the usual hydrogen–oxygen kinetic mechanisms are the reactions involving electronically excited molecules. These may appear, for example, behind a strong oblique shock wave in a shock-induced ignition configuration. Table 6.3 shows several of the reactions that include O_2 in the $^1\Delta_g$ electronic state (which is located at 0.98 eV above the ground state;

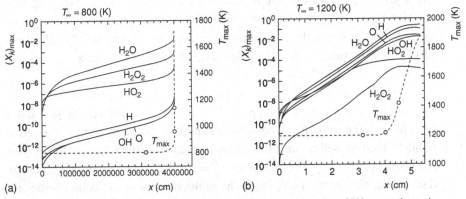

(a)   (b)

Figure 6.16. Axial distribution of species maximum mole fraction $(X_k)_{max}$ and maximum temperature T_{max}, originating from a splitter plate for hydrogen–air chemistry (from Nishioka and Law, 1997): (a) the low initial temperature ($T_\infty = 800$ K) indicates the dominance of HO_2 production over the chain-branching reactions producing O, H, and OH; and (b) the situation is reversed at higher temperatures ($T_\infty = 1200$ K).

Herzberg, 1989) included by Starik and Titova (2001) in their 45 hydrogen–oxygen-reaction mechanism. This reaction mechanism includes ozone reactions in addition to the reactions in Jachimowski's mechanism. The reaction numbers in the list included in Table 6.3 follow the original numbering in Starik and Titova's mechanism to indicate that only selected reactions are listed here. These are the reactions in which electronically excited oxygen participates.

Clearly, for the reactions involving the vibrational and electronic excited O_2 molecules, the exothermic reaction rates are increased by the amount of the activation energy that leads to the molecule excitation. Only part of the calculated excitation energy should be included in the activation energy because part of the electronic or vibration energy is dissipated before fully contributing to the oxygen molecule excitation. This correction is accounted for by Starik and Titova (2001) by including a *coefficient of electronic* or *vibrational energy utilization* α_q, defined as

$$\alpha_q = \frac{E_{aq}^+}{E_{aq}^+ + E_{aq}^-},\qquad(6.31)$$

where E_{aq}^+ is the activation energy of the qth chemical reaction for the destruction of the excited molecule and E_{aq}^- is the activation energy for the reaction leading to the formation of the activated molecule. With this additional energy, the reaction rate of the qth chemical reaction is modified to

$$k_q = A_q T^n \exp\left(-E_{aq} + \alpha_q E_{\text{excitation}_q}/RT\right),\qquad(6.32)$$

where T is the translational temperature.

Reactions 31, 32, and 43 in Table 6.3 include a term, q_a, that corrects the preexponential constant A. This term arises from the degeneracy of the electronic ground state for oxygen, which includes the $^3\Sigma_g^-$, $^1\Delta_g$, and the $^1\Sigma_g^+$ levels (Herzberg, 1989), of which only the $^1\Delta_g$ electronic level was included. For the reactions in which the correction factor did not appear, Starik and Titova (2001) used the reaction rates calculated by previous researchers.

The result of including oxygen in the electronic excited state accelerates the chemical reactions, as expected. For the ignition process in the low-temperature range, which is supported by chain-branching reactions 2 and 3 in Table 6.2, the initiation time is determined by the formation of the active radicals. Ignition will occur if this chemical characteristic time τ_i is lower than the characteristic time for diffusion, $\tau_{ik}^D = l^2/D_{ik}$, where l is a characteristic distance for the reaction and D_{ik} is the multicomponent diffusion coefficient. The diffusion time for the hydrogen–oxygen system is the diffusion time for H because reaction 2 is considered the most important for ignition (Balakrishnan et al., 1995). Figure 6.17, after Starik and Titova (2001), shows a calculation of the diffusion time for H and the ignition time for a premixed, stoichiometric hydrogen–oxygen system and for mixtures in which excited O_2 molecules are

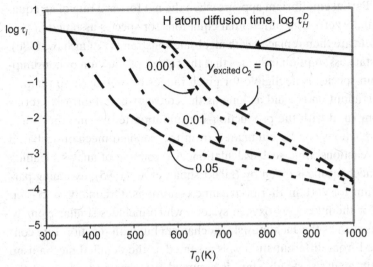

Figure 6.17. Calculated ignition times for premixed hydrogen–oxygen depending on the amount of electronically excited O_2 in the mixture. The dotted curve indicates the hydrogen atom diffusion time for premixed stoichiometric hydrogen–oxygen mixtures at $P_0 = 10\,kPa$. The following curves indicate the induction time calculated for selected proportions of electronically excited O_2, $y_{\text{excited }O_2}$, in the total content of oxygen molecules. The induction time drops monotonically as the proportion of electronically excited O_2 molecules increases.

present in selected concentrations. Without the presence of the electronically excited O_2, the diffusion time and the initiation time intersect, for the stoichiometric mixture at the selected 10-kPa pressure, at the self-ignition temperature $T = 580\,K$. This temperature drops as the electronically excited O_2 molecules are added even in proportions as small as 0.1% of the total O_2 molecules and continues to drop monotonically as the proportion of electronically excited O_2 molecules increases. As the initial mixture temperature increases, the overall effect of electronically excited O_2 decreases. The main contribution of O_2 $(^1\Delta_g)$ is through reactions 38 and 39, which are both chain branching with faster reaction rates than their counterparts with O_2 in the ground state.

In practice, the appearance of excited oxygen molecules may be due to the presence of a strong shock wave, anchored, for example, to the vehicle inlet capture; the activated molecule may also appear during initiation with an electric discharge. The result is a significant reduction of the induction time, thus allowing sustained combustion at low temperatures and an overall reduction of the combustion zone.

6.4.1.1 Reduced Mechanisms for Hydrogen–Air Combustion

A number of chemical reactions can be eliminated from the detailed reaction mechanism to reduce the stiffness of the set of equations, assuming that particular reactions are in partial equilibrium or under species steady-state assumption. Partial equilibrium appears when the net forward rate of an equation is essentially zero. The differential equations for species assumed in partial equilibrium are then replaced with algebraic constraints (Ramshaw, 1980). The steady-state assumption indicates that the rate of production or consumption of certain species is negligible for given ranges of pressure and temperature. This assumption is valid as long as the convective and diffusive terms are small compared with the production–consumption terms, and thus some elementary reactions can be eliminated from the detailed mechanism based on separate reactions' time-scale calculation. An example of analysis leading to reduced mechanisms is given by Balakrishnan et al. (1995), assuming partial equilibrium for certain thermodynamic conditions. Originally, a 21-step mechanism for the hydrogen–oxygen system, which includes similar elementary reactions such as the Jachimowski mechanism but with reaction-rate constants adopted from different studies, is assumed as the detailed mechanism. Then, through analysis, steady state is assumed for certain species and the mechanism is reduced. In the low-pressure low-temperature range, HO_2 is a stable molecule that can be assumed at steady state in the flame region with a net source term $|\omega_f - \omega_b/\omega_f|_{HO_2} \leq 10^{-7}$ (Balakrishnan et al., 1995). With HO_2 at steady state, the formation of H_2O_2 and the reactions that involve this species do not materialize. A four-step reaction mechanism is then suggested, including

$$H + O_2 \rightleftarrows OH + O,$$
$$O + H_2 + M \rightleftarrows H + OH + M,$$
$$OH + H_2 \rightleftarrows H + H_2O, \tag{6.33}$$
$$H + H + M \rightleftarrows H_2 + M,$$

with reaction rates determined from algebraic combinations of elementary reactions from the detailed mechanism (the reaction rates and third-body efficiencies are listed in Balakrishnan et al., 1995). Once the reduced reaction mechanism is verified against the results produced by the detailed mechanism, it can be used successfully within certain limitations. These include the selection of temperature, pressure, and strain-rate range when some reactions can be assumed at steady state with a reasonable degree of accuracy. The four-reaction mechanism shown by reactions (6.33) has been successfully tested in the main reaction zones or for high-temperature ignition. For other thermodynamic conditions, such as the wings of diffusion flames, the same reduced reaction mechanism may not accurately reproduce the radical concentrations, and different reactions have to be selected for the reduced mechanism. In that regard, there are still large differences among the various reactions mechanisms from one study to another. For example, a recent compilation of forward reaction rates of the chain-branching $O + H_2 \rightarrow OH + H$ reaction by

Javoy et al. (2003) indicated that the results from different studies may exhibit more than 20% variation in the 2000–2500 K range. Balakrishnan et al. (1995) showed that, depending on the temperature, strain rate, and mixture fraction regimes, reduced mechanisms can result in even larger variations in radical concentration. Therefore the reaction mechanisms used in any particular scramjet analysis must be chosen according to the thermodynamic and combustion regime expected in the volume of flow of interest.

Finally, the one-step hydrogen–oxygen reaction,

$$H_2 + \frac{1}{2}O_2 \rightarrow H_2O, \tag{6.34}$$

covers the complete combustion process but is highly inaccurate for describing the flame structure and the heat release under the thermodynamic conditions met in the scramjet engine. Considering the computational capabilities available today, there are few cases that justify using the one-step reaction for scramjet-flow-related analyses.

6.4.2 Reaction Mechanisms for Hydrocarbons

The chemistry of hydrocarbon systems includes significantly more equations than for hydrogen combustion, which is, in fact, a subsystem of the former. In general, hydrocarbon fuels are complex systems with many carbons in the molecules; detailed reaction mechanisms can be extensive and difficult to manage computationally. For example, the mechanism for NO_x production in methane flames proposed by Li and Williams (1999) contains 177 reactions; the reaction mechanism for C_1–C_3 systems proposed by Leung and Linstedt (1995) includes 451 reactions. Reduced reaction mechanisms are therefore particularly useful for hydrocarbon reactions.

The oxidation of large hydrocarbon compounds is assumed to contain consecutive steps that lead to the formation of the stable methyl and ethyl groups, which are then oxidized through reaction mechanisms that have been substantially verified (Warnatz et al., 1996). A general scheme that contains successive hydrocarbon decompositions reactions would include the following steps (Chomiak, 1990):

1. Free-radical (H, O, or OH) attack on the hydrocarbon breaks a C–H bond and creates an activated radical $\dot{R}$:

$$H, O, OH + RH \rightarrow H_2, OH, H_2O + \dot{R}.$$

2. Decomposition of the radical $\dot{R}$ by a C–C scission forming a lower hydrocarbon:

$$\dot{R} \rightarrow RH + CH_3,$$

or a reaction of $\dot{R}$ with O_2 to form a lower hydrocarbon or other oxygenated products:

$$\dot{R} + O_2 \rightarrow RH + HO_2,$$
$$\dot{R} + O_2 \rightarrow \dot{R}O_2.$$

3. Decomposition of $\dot{R}O_2$ via isomerization to form $RO\dot{O}H$:

$$\dot{R}O_2 \rightarrow RO\dot{O}H,$$

or to form lower hydrocarbons:

$$\dot{R}O_2 \rightarrow RH + OH, H, \text{ etc.}$$

4. Chain-branching decomposition of $RO\dot{O}H$:

$$RO\dot{O}H \rightarrow \dot{R}O + OH.$$

5. Continuous reactions leading to lower hydrocarbon systems.

It should be noted that this generalization is in itself a substantial simplification because recombination reactions can be present that lead to the formation of hydrocarbon systems that are even higher than those in the initial mixtures. This includes not only linear higher systems but also aromatic and polyaromatic rings that, as precursors to soot formation, have been observed in many flames of relatively low-hydrocarbon systems, in particular in rich flames with $\phi > 1$. Several examples of reduced hydrocarbon oxidation mechanisms are subsequently discussed briefly.

REDUCED REACTION MECHANISM FOR METHANE. As an example, a reduced reaction mechanism for methane (Williams, 1999) is

$$\begin{aligned}
CH_4 + H &\rightarrow CH_3 + H_2, \\
CH_4 + OH &\rightarrow CH_3 + H_2O, \\
CH_3 + O &\rightarrow CH_2O + H, \\
CH_3 + H + M &\rightarrow CH_4 + M, \\
CH_2O + H &\rightarrow CHO + H_2, \\
CH_2O + OH &\rightarrow CHO + H_2O.
\end{aligned} \qquad (6.35)$$

These reactions include O, H, and OH radicals that must be formed from additional elementary steps involving hydrogen and oxygen chemistry, which has been discussed before. In the mechanism given by reactions (6.35), the first two reactions correspond to the H atom abstraction described in the general scheme suggested by Chomiak (1990), followed by methyl, formaldehyde, and formyl consumption. The termolecular reaction listed in reactions (6.35) is encountered in low-pressure systems and becomes a bimolecular reaction in high pressures. The reaction rate must then be adjusted to correspond to the pressure conditions (Williams, 1999). The reaction rates for the other reactions derive from steady-state assumptions and are discussed in detail by Williams (1998).

A further simplification is the four-step reaction mechanism that includes consumption of methane to CO followed by the water–gas shift reaction and two other reactions that represent O_2 consumption leading to formation of H and the possible recombination of H:

$$CH_4 + 2H + H_2O \rightarrow CO + 4H_2,$$
$$CO + H_2O \rightarrow CO_2 + H_2, \qquad (6.36)$$
$$3H_2 + O_2 \rightarrow 2H + 2H_2O,$$
$$2H \rightarrow H_2.$$

In fact, for drastically simplified reaction mechanisms for alkanes, the first reaction in the list shown in reactions (6.36) could be replaced with a generic fuel-consumption step:

$$C_nH_{2n+2} + \alpha H + nH_2O \rightarrow nCO + \left(2n + 1 + \frac{1}{2}\alpha\right)H_2, \qquad (6.37)$$

where α is determined from the hydrogen atom conservation. As the hydrocarbon compounds become more complex, there are additional paths for their decomposition along with the hydrogen abstraction characteristic for methane decompositions; they appear via the C–C bond break characteristic of step 3 in Chomiak's (1990) model.

A MECHANISM FOR ETHYLENE IGNITION. Ethylene chemistry is of interest for scramjet-related studies (Kay et al., 1990) because it is an important constituent in the thermal decomposition of several liquid-hydrocarbon compounds such as JP-10. The studies by Varatharajan and Williams (2002a, 2002b) proposed an extended reaction mechanism for ethylene ignition with 148 elementary reactions. The reaction rates in this mechanism are based on measured ignition times in a number of shock-tube studies spreading over three decades. The applicability of this mechanism and of the reduced mechanisms derived from it lies within 0.1 to 10 atm, 1000–2500 K, and equivalence ratios of 0.5–2, which cover a broad range of conditions met in the scramjet engine.

From this mechanism's predictions of temperature and species concentration time history and using steady-state and partial-equilibrium approximations, we find that further simplifications reduce the mechanism first to 38 reactions and then to only the 12 elementary reactions for ethylene ignition:

$$H + O_2 \rightarrow pOH + pO + (1 - p) HO_2,$$
$$H_2O_2 \rightarrow 2OH,$$
$$C_2H_4 + O_2 \rightarrow C_2H_3 + HO_2,$$
$$C_2H_4 + OH \rightarrow C_2H_3 + H_2O,$$
$$C_2H_4 + O \rightarrow qCH_3 + qCHO + (1 - q) CH_2CHO + (1 - q) H,$$
$$C_2H_4 + 2HO_2 \rightarrow CH_3 + CO + H_2O_2 + OH, \tag{6.38}$$
$$C_2H_4 + H \rightarrow rC_2H_3 + rH_2 + (1 - r) C_2H_5,$$
$$C_2H_3 + O_2 \rightarrow sCH_2O + sCHO + (1 - s) CH_2CHO + (1 - s) O,$$
$$CH_3 + O_2 \rightarrow CH_2O + OH,$$
$$CH_2CHO \rightarrow CH_2CO + H,$$
$$C_2H_5 + tO_2 \rightarrow C_2H_4 + tHO_2 + (1 - t) H,$$
$$CHO + (1 - u) O_2 \rightarrow CO + uH + (1 - u) HO_2.$$

The nondimensional parameters p through u appearing in the reactions (6.38) depend on the reaction rates discussed in detail by Varatharajan and Williams (2002b); these reaction rates derive from the 38 reaction mechanisms. Further analysis of the ignition process at low and high temperature leads to the formulation of a seven-step reaction mechanism that was shown to predict ignition times that differ from the detailed mechanism by a factor of two.

N-HEPTANE–AIR OXIDATION. With C_7 and higher hydrocarbon systems, the reaction mechanisms become considerably more complex. There have been several detailed studies for n-heptane oxidation (for example, Linstedt and Maurice, 1995; Ranzi et al., 1995; Slavinskaya and Haidn, 2003), and the extent of the reaction mechanisms suggested by these studies is too complex to be included here. The systematic approach taken by Curran et al. (1998) formulated a detailed model that emphasizes classes of reactions that are representative of temperature regimes and follows with analyses of types of reactions. The complete reaction mechanism suggested by the analysis of Curran et al. (1998) for n-heptane oxidation includes 2450 elementary reactions with 550 species and was verified for pressures ranging between 1 and 42 atm, temperatures from 550 to 1700 K, and equivalence ratios between 0.3 and 1.5; Slavinskaya and Haidn (2003) suggested a reduced model for both n-heptane and iso-octane (or mixtures of the two) in air with 1006 reactions and 134 species validated against pyrolysis and ignition experimental data for the range of pressures from 6 to 44 atm, temperatures between 650 and 1200 K, and equivalence ratios between 0.5 and 2.

The initiation of the mechanism starts in cases of both high- and low-temperature regimes by hydrogen abstraction, resulting in four possible alkyl radicals. At high temperatures these radicals undergo a β-scission in the way described by the generic model suggested by Chomiak (1990) and Warnatz et al. (1996). At lower temperatures, however, the alkyl radicals are modified by the addition of O_2, and hence each mechanism proceeds with the hydrocarbon compound decomposition through different classes of reactions. The major classes of reactions for n-heptane oxidation identified by Curran et al. (1998) are subsequently given and represent, as the authors suggest, a basis for the formulation of more complex mechanisms adapted for specific applications (here R and R' represent alkyl radicals and structures, and Q represents alkene and alkene structures):

1. Unimolecular fuel decomposition.
2. H atom abstraction from the fuel.
3. Alkyl radical decomposition.
4. Alkyl radical $+ O_2$ to produce olefin $+ HO_2$ directly.
5. Alkyl radical isomerization.
6. Abstraction reactions from olefin by OH, H, O, and CH_3.
7. Addition of radical species to olefin.
8. Alkenyl radical decomposition.
9. Olefin decomposition.
10. Addition of alkyl radicals to O_2.
11. $R + R'O_2 = RO + R'O$.
12. $RO_2 \rightarrow ROOH$.
13. $RO_2 + HO_2 = RO_2H + O_2$.
14. $RO_2 + H_2O_2 = RO_2H + HO_2$.
15. $RO_2 + CH_3O_2 = RO + CH_3 + O_2$.
16. $RO_2 + R'O_2 = RO + R'O + O_2$.
17. $RO_2H = RO + OH$.
18. RO decomposition.
19. $QOOH = QO + OH$.
20. $QOOH = $ olefin $+ HO_2$.
21. $QOOH = $ olefin $+$ carbonyl $+ OH$.
22. Addition of $QOOH$ to O_2.
23. Isomerization of O_2QOOH and formation of ketohydroperoxide and OH.
24. Decomposition of ketohydroperoxide to form oxygenated radical species and OH.
25. Cyclic ether reactions with OH and HO_2.

The model and the sensitivity analyses indicate that accurate modeling of the hydrocarbon decomposition rests with the careful description of the chain-branching reactions and the kinetic processes that compete with them. The extensive analysis results of Curran et al. (1998), in comparison with shock-tube data, indicated that although good agreement was generally noticed, the reactions paths and the reaction rates could still be improved by use of the available experimental information.

IGNITION OF JP-10. The interest in JP-10 was generated because it is a high-energy–density liquid compound with applications for ramjet–scramjet engines. It is a single compound fuel with a known chemical formula ($C_{10}H_{16}$) that makes it easier to analyze than other practical fuels. In general, fuels for propulsion applications are blends of different hydrocarbons, and the compounds involved can be large systems that cannot be easily modeled. Therefore attempts to describe reaction mechanisms for these blends have modeled simplified systems with an estimated carbon number representing an average of the blend (Linstedt and Maurice, 2000).

Li et al. (2001) offered an extended chemical mechanism consisting of 174 elementary steps with 36 chemical species in which 13 reactions describe JP-10 decomposition to C_3–C_5 systems. This reaction mechanism was developed to predict ignition times for JP-10 within the temperatures range of 1000–2500 K, pressures between 1 and 100 atm, and equivalence ratios between 0.5 and 2.0; the predictions based on this mechanism were found to be in good agreement with data from JP-10 shock-tube experiments. Based on the study by Li et al. (2001), the JP-10 concentration continuously decreases during ignition and the radical concentrations and temperature remain relatively constant; they begin to increase only after the JP-10 concentration has fallen below the radical concentrations. This indicates that JP-10 is a strong sink for radicals so that appreciable branching can begin only after the JP-10 concentration becomes negligible in comparison with concentrations of major radicals such as OH and H.

When steady-state approximations were used for a number of species during the ignition process, a further reduction of this mechanism was suggested to include only three decomposition reactions for JP-10 in addition to the ignition chemistry for methane, acetylene, and ethylene for which existing reaction mechanisms can be used.

6.4.3 Summary

If hydrogen is the selected fuel for the scramjet application, detailed reaction mechanisms for ignition and the combustion zone are sufficiently well developed to provide accurate species concentrations and temperature temporal descriptions. These mechanisms include a sufficiently small number of reactions to be manageable with today's computational capabilities. Hydrocarbon fuels continue to represent a challenge. As the number of carbons in the fuel molecule increases, the number of chemical reactions required for describing the combustion process accurately becomes exceedingly large. Detailed mechanisms for even C_7 hydrocarbon fuels reach thousands of elementary reactions, yet most of the candidate fuels include larger hydrocarbon systems. These mechanisms must be validated against experimental data over a range of initial pressures, temperatures, and equivalence ratios. Reduced mechanisms are needed, but they can be developed only once a detailed mechanism has been established and verified against experimental data, because the reduced mechanisms can be only an approximation of the detail mechanism. In that regard, despite significant analytical efforts made to date – of which only several studies have been mentioned here – much work remains ahead.

6.5 Flame Stability

Flameholding requires achieving a balance between the flame propagation speed and the fluid velocity. Because the fluid velocity exceeds the flame speed in supersonic combustion applications, the flameholding issue is solved by the generation of some sort of recirculation region that ensures sufficient residence time so that the processes involved – fuel–air mixing, ignition and chemical-reaction propagation – can take place to completion. These processes are determined by local conditions of gas composition, temperature, and velocity and are substantially different in nonpremixed cases, such as those encountered in most practical applications, than in premixed cases, which are easier to predict and analyze.

A substantial database of flame stability exists for premixed gases (Huellmantel et al., 1957; Ozawa, 1971; Ogorodnikov et al., 1998) from which stability limits for rich and lean flames have been obtained for a number of fuel–air systems. The stability limit is usually cast in terms of a flameholding boundary on an equivalence ratio versus a stability parameter plane. The stability parameter depends, in general, on the flow velocity, temperature, size, and shape of the flameholder and has received various formulations in different studies, from empirical formulations to expressions that reflect global Damkhöler numbers.

In the case of non-premixed gases the determination of stability limits is less straightforward, mostly because of the nonhomogeneity of the parameters in the recirculation region behind the flameholder. It is difficult to estimate the spatial species concentration and temperature distribution in the recirculation regions of these flows because of the presence of large gradients and the complex 3D flow structure. These difficulties are compounded by the uncertainty in the shape of the recirculation region, which depends on the amount of heat release that, in turn, is dictated by the local mixing and combustion efficiencies.

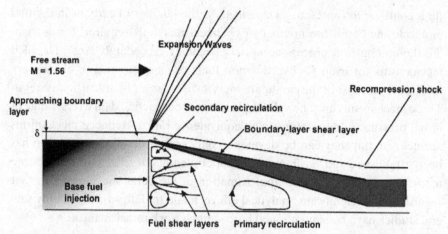

Figure 6.18. A 2D recirculation region flow-field schematic. The boundary layer
formed upstream of the step is pushed by the expansion toward the test section wall.
A shear layer forming between the boundary layer and the recirculation region brings
fresh air in the region. The fuel is injected into the recirculation region, and a barrel
shock forms at each injection orifice with shear layers in which the flame is initiated
following mixing and heat exchange between the hot gases in the region. A primary
recirculation of gases exists that engulfs the recirculation region with additional smaller
recirculations present. Additional 3D flow patterns, not shown in the figure, exist.

The following discussion focuses on the characteristics of the flow field in the
region of a recirculation region with implications for the flameholding analysis
and modeling.

6.5.1 Recirculation-Region Flow Field

A simple recirculation region is the rearward-facing step shown schematically
in Fig. 6.18. The rearward-facing step has been among the early solutions for
flameholding in supersonic flows and continues to exist in combination with
other geometrical configurations in most currently proposed flameholders.
Among the advantages presented by the rearward-facing step are (i) the good
separation between the increased pressure that is due to combustion in its base
and the upstream incoming flow and (ii) the absence of intrusive devices that
may generate stagnation pressure loss and require internal cooling.

Although the flow is highly 3D even for a 2D flameholder, because of
the effects of the side walls that are generally present, only a 2D description
is included here, for simplicity. The boundary layer formed upstream of the
step arrives at the recirculation region and is pushed by the expansion toward
the wall. A shear layer forms between this boundary layer and the recirculation region, bringing fresh air into the region. A primary recirculation of gases
exists that engulfs the recirculation region, as is indicated in the figure, with
additional smaller recirculations present close to the walls. If fuel is injected

into this region, it is usually an underexpanded jet, as shown in the figure, and a barrel shock forms at each injection orifice. Other forms of refueling of the recirculation region exist, for example, injection from the wall downstream of the reattachment point or upstream fueling of the boundary layer. In both these cases, fuel and fresh air arrive into the recirculation region by mass transfer through the shear layer. Any species concentration within this shear layer can have any value, at a given instance, complicating the prediction of the gas exchange (Dimotakis, 1991). In the case shown in the figure, shear layers develop at the jet–gas boundary, in which the flame is initiated following mixing and heat exchange between the hot gases in the region. The fuel jet may transition from underexpanded to sonic and even subsonic, with significant implications for the development of the local shear layers, and hence for mixing and local heat release. In turn, this heat release modifies the structure of the recirculation region, further affecting mixing.

Morrison et al. (1997) offered an empirical estimate of fuel–air mass exchanges in the recirculation region formed behind a flameholder such as the one in Fig. 6.18 by correlating the size of the recirculation region and the estimated residence time to obtain a local equivalence ratio. Along with other local parameters responsible for flame stability, such as temperature and pressure, a stability parameter of the type suggested by Ozawa (1971) for premixed gases was proposed. With the observation that the recirculation region remains subsonic, Morrison et al. (1997) suggest that stability parameters derived for subsonic, premixed gases can be applied to a certain extent for non-premixed flows, as well. The underlying assumption is that mixing of the fuel injected into the recirculation region is complete and uniform.

The dominant effect in the development of shear layers between the fuel jets and the surrounding gases is indicated in Fig. 6.19, which shows the pressure change in the recirculation region as the fuel flow is reduced (Ortwerth et al., 1999). In the absence of local information, the equivalence ratio in this plot, Φ_{Base}, is based on the entire mass of air flowing through the device, P_{Base} is the pressure measured in the base of the rearward-facing step, and P_s is the static pressure upstream of the step used as a normalizing factor. The relation to the local equivalence ratio will be discussed in a subsequent section. At high equivalence ratios, large amounts of fuel are expected to leave the recirculation region and continue to burn in the high-speed flow region (Wright and Zukoski, 1960). Below $\Phi_{Base} = 0.07$, the fuel flow rate is sufficiently low to mix and burn within the recirculation region without flame propagation as indicated by the "kink" in the curve. This limit also indicates the domain below which the recirculation-region length remains unaffected by combustion.

Therefore, based on these, mostly qualitative, observations, it can be concluded that stability parameters based on conditions external to the bluff body, e.g., the upstream conditions, in particular velocity and stagnation

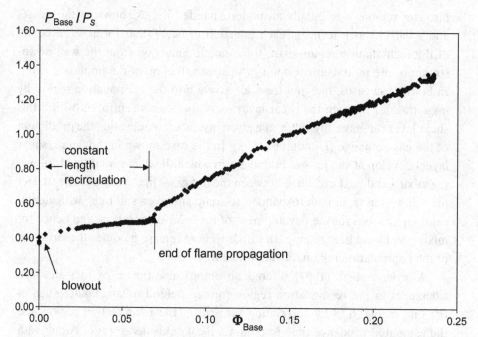

Figure 6.19. Normalized base pressure rise vs. a global equivalence ratio. The "kink" in the curve indicates the limit of flame propagation through the recirculation region. At lower Φ_{Base} the recirculation region length remains constant (Ortwerth et al., 1999).

temperature with assumptions of fixed recirculation-region length (Ozawa, 1971), may introduce large uncertainties in the analyses of non-premixed flows.

Figure 6.19 helps to identify a distinction that should be made between two conditions of importance in the operation of a supersonic combustion chamber: (i) a boundary of flame spreading, represented by the boundary beyond which the flame extends beyond the recirculation region and (ii) a boundary of residual flame (Ogorodnikov et al., 1998) below which the flame is lost altogether. The beginning of pressure increase in the combustion chamber identifies the first condition, also called *blowoff*, as heat is continuously added to the flow. The sharp drop in temperature measured in the recirculation region identifies the latter, also known as *blowout*.

6.5.2 Recirculation-Region Temperature

There are sharp gradients of temperature in the recirculation region, and they change location as the heat release and the geometric shape of the recirculation region change (Ogorodnikov et al., 1998). Thus the assumption of a uniform temperature in the recirculation region based, for example, on a single point measurement is only indicative and may not represent correct average or global properties. For example, Figure 6.20 (Owens et al., 1997) shows a

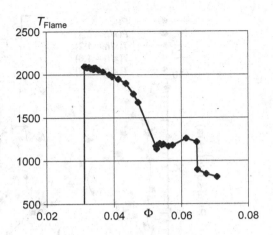

Figure 6.20. Temperature measured with a thin, B-type thermocouple (Owens et al., 1997) as a function of the global equivalence ratio Φ indicates values close to adiabatic flame temperatures for stoichiometric mixtures.

measurement taken at $1/2H$ from the step in an axial direction and $1/2H$ above the chamber wall for a configuration that includes distributed fuel injection from base orifices. This type of injection achieves a rapid 2D fuel distribution at low equivalence ratios, resulting in good flameholding characteristics. The temperature T_{Flame} in Fig. 6.20 is related to the overall equivalence ratio Φ, during fuel throttling from large to low Φ. This equivalence ratio is based on the total amount of fuel injected and the total amount of air traveling through the combustion chamber and therefore indicates significantly lower values than are actually present in the recirculation region. As can be seen in the figure, temperatures close to adiabatic flame values for stoichiometric mixtures are noted, reflecting the substantial difference in the local equivalence ratio experienced and the values estimated based on global conditions. The sharp drop in the indicated temperature, as the equivalence ratio is reduced, corresponds to the physical destruction of the thermocouple as the local temperature approaches stoichiometric values.

Simultaneous measurements at multiple locations in the recirculation region (Owens et al., 1997) indicate the substantial gradients present in the small region occupied by the recirculation region. The study presented by Owens et al. (1998) investigated the blowout limits in a non-premixed hydrogen–air system stabilized behind a rearward-facing step with supersonic injection from orifices placed in the base of the recirculation region. The resulting reacting flow was thus embedded in a supersonic airflow with parameters upstream of the flameholder characterized by Mach 1.8 and stagnation temperatures in the range of 600–1000 K. The combustor operated in the pseudoshock mode, as first described by Shchetinkov (1973) and later by Heiser and Pratt (1994).

Figure 6.21, taken from Owens et al. (1998), shows these temperature changes were recorded simultaneously at three different locations, indicated in the figure in terms of step height h as the amount of fuel injected was changed.

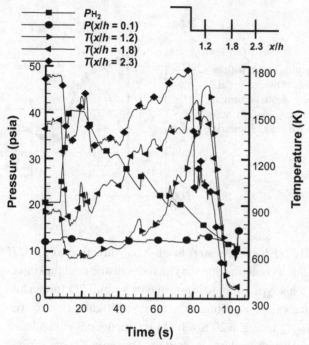

Figure 6.21. Although the local pressure P remains constant, therefore indicating a fixed-size recirculation region, the temperatures show large gradients in time and space.

Also included in the figure is the change in the fuel-injection pressure P_{H_2} as a measure of fuel throttling. Although the local pressure P remains constant, therefore indicating a fixed-size recirculation region, the temperatures, measured at 1.2, 1.8, and 2.3 step heights from the step, show large gradients in time and space. Furthermore, the increase in local temperature as the overall injected fuel flow is reduced indicates that, locally, the mixture is fuel rich, in contrast to the conclusion that would be drawn based on the global combustion chamber equivalence ratio. Therefore a direct correlation between the flameholding region gas composition and the global parameters is not straightforwardly available.

6.5.3 Local Equivalence Ratio Analysis

The equivalence ratio in the stability analysis that follows for the data provided by Owens et al. (1998) is based on the total incoming airflow. From the analysis suggested by Morrison et al. (1997), the estimated air replenishment flow into the recirculation region is about 3.7% of the total device's airflow for simple, rearward-facing recirculation regions, as discussed here, when the recirculation length is $5h$, and 1% when the recirculation region shrinks to $1h$ as the fuel flow and consequently the heat released are reduced. Additional analysis can be performed based on shear-layer development under the assumption that the transfer of fresh air and burned gases into and from the recirculation region is controlled by the development of the shear layer at the recirculation-region boundary. Pitz and Daily (1983) indicate that the rate of growth of shear layers for a given duct expansion is insensitive to effects of combustion in the shear layer itself and remains constant to about $\delta/h = 0.28$. Correcting for the compressibility effects (Dimotakis, 1991) by means of Mach and Reynolds numbers at the experimental conditions of the study by Owens et al. (1998), from which the data in Fig. 6.21 were taken, i.e., M = 1.8 and Re = 1.2×10^6, the following correction factors for the shear-layer development are found:

$$f_{Re} = 0.75; \quad f_M = 0.4. \tag{6.39}$$

With these data, the air mass flow in the shear layer at reattachment, which is responsible for replenishment of the recirculation region, can be calculated. Assuming a reattachment length of $5h$ (Morrison et al., 1997) and with the velocity and density ratios estimated from the experimental data as $r = 0.57$ and $s = 0.25$, respectively, the equivalence ratio correction becomes

$$\phi_{factor} = 1/2(\delta/h)(x_r/h) f_{Re} \, f_M rs = 0.03, \tag{6.40}$$

with 1/2 reflecting the symmetry of the test section. This equivalence ratio estimate is in remarkably good agreement with the observed peak recirculation-zone temperature, which indicates local stoichiometry.

6.5.4 Recirculation-Region Composition Analysis

An example of gas composition in the recirculation-region analysis is offered by Thakur and Segal (2003), who measured through mass-spectroscopic sampling the distribution of an inert injectant in the recirculation region formed behind a sudden expansion in a supersonic flow. Sampling from the wall along a region extending beyond the physical size of the recirculation region and transverse to the recirculation region at selected locations confirmed that locally the injectant concentration is substantially larger than estimated based on global parameters. Figure 6.22 shows the argon mass-distribution samples from the wall at locations extending beyond $3h$, at a flow condition that resulted in a recirculation region length of $<2h$. The figure indicates that the local argon mass fraction is found to be 3–8 times larger than the estimate based on global conditions in both experimental cases, which included a higher and a lower injection pressure. It is interesting to note that, although some variation exists in the axial direction with a drop toward the end of the recirculation region and a slight increase beyond it, the differences are not significant, indicating that sufficient argon propagation beyond the recirculation region took place. The results of Thakur and Segal (2003) show that sampling

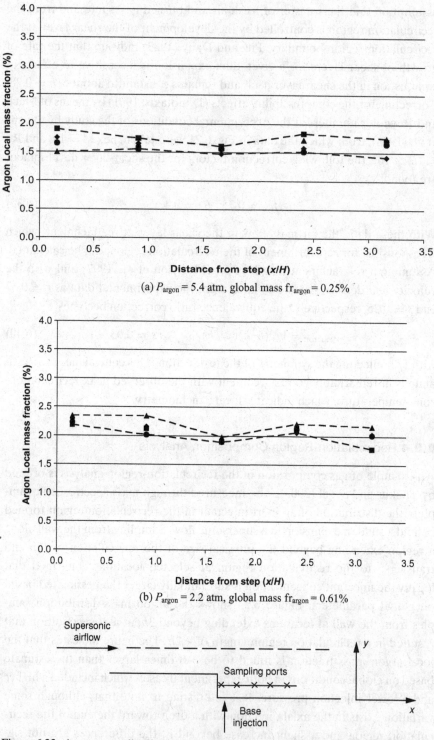

(a) P_{argon} = 5.4 atm, global mass fr$_{argon}$ = 0.25%

(b) P_{argon} = 2.2 atm, global mass fr$_{argon}$ = 0.61%

Figure 6.22. Argon mass distribution obtained with argon injection in the recirculation region at two different pressures. A slight decrease toward x/H = 1.5, which marks the end of the recirculation region, is noted, followed by an increase in the injectant concentration beyond it.

in a transverse direction to the flow in the recirculation region indicates a variation of local concentration caused by a clearly 3D flow and an inflow argon concentration exceeding the wall sampling by 20%–30%, depending on the injection pressure.

6.5.5 Stability Parameter Formulations

It has been suggested (Morrison et al., 1997) that, because the flow remains subsonic in the recirculation region, flame stability parameters obtained from subsonic flows can be applied to subsonic flameholder flow regions that are embedded in a supersonic flow. The local equivalence ratio ambiguity can then be solved by estimating the amount of fresh air that enters the recirculation region and treating it as homogeneous. This includes the underlying assumption that mixing is fast and uniform, which is reasonable at low equivalence ratios. Citing a large database of previous studies, Ozawa (1971) formulated an empirical equation that relates the amount of air mass flow into the recirculation region to the total mass flow and the geometrical shape at the flameholder. A stability parameter is then defined by Ozawa for premixed gases, taking the following formulation:

$$SP = \frac{V}{d} f_d \frac{1}{P} \left(\frac{1000}{T_0} \right)^{1.5}, \qquad (6.41)$$

where SP is the stability parameter, V is the air velocity arriving at the flameholder, d is the physical size of the flameholder f_d, a factor depending on the shape of the flameholder, P is the gas static pressure, and T_0 is the stagnation temperature upstream of the flameholder. The thermodynamic parameters involved in the stability parameter equation have a clear and intuitively expected effect on flame stability. On a plot of equivalence ratio Φ versus the stability parameter SP, a curve called the *stability loop* separates the region of stable flames from the region when the flames blow out. These curves have a maximum at stoichiometric conditions.

Other stability parameter formulations have been suggested for flameholders embedded in supersonic flows with parameters measured both upstream and in the recirculation region. For example, Wright and Zukoski (1960) suggested a stability criterion for cavity flameholders of length L that depends both on the local pressure in the cavity P_l and the upstream parameters, i.e., velocity V and stagnation temperature T_0. This parameter is then plotted as a function of the local equivalence ratio, measured in the cavity, as

$$K_s = \frac{V}{P_l^{1.45} T_0^2 L}. \qquad (6.42)$$

Other formulations have been suggested based on an exponential temperature dependence (Strokin and Grachov, 1997) as follows:

$$K_{da} = \frac{V\frac{dF}{dx}}{P\exp^{(-1.12^{T_0/1000})}},\tag{6.43}$$

where dF/dx is the local combustion chamber cross-sectional area change.

A stability criterion that directly describes a local, global Damköhler number was formulated by Ogorodnikov et al. (1998) as

$$K_{da'} = \frac{V}{P_{\text{inj}}\exp^{(-1000/T_{\text{local}})}},\tag{6.44}$$

where P_{inj} is the local injection pressure and T_{local} is the local flame temperature measured in the recirculation region. The fuel-injection pressure is used in this expression to indicate the dependence of the mixing processes on the local shear-layer development. The shortcomings of the formulation appear in the residence time, introduced by use of the air velocity immediately upstream of the flameholder and an assumed constant recirculation-region length. Nevertheless, although largely simplifying the complex processes that take place in the flameholding region, this stability criterion includes local flow parameters responsible for mixing and combustion and captures the major physical processes involved.

It is interesting to pay attention to the effect the development of the shear layer at the flameholder has on the flame stability. Figure 6.23 shows the global equivalence ratio at blowout Φ_b versus a shear-layer parameter responsible for mixing (Ortwerth et al., 1999), where V_{max} and V_{min} are the velocities and s is the density ratio on the two sides of the shear layer under the experimental conditions described by Ortwerth et al. (1999). Two types of recirculation regions were used in this study with essentially the same flameholding results, which are shown in the figure at two different air stagnation temperatures. There appear to be separated regimes dictated by the changes in the shear-layer development, which is responsible for changes in the mixing length and hence the flame stability. A vertical boundary appears to form in all cases in which the fuel is injected in large quantities; thus fuel in large amounts and at high velocities may leave the recirculation region without participating in the local combustion. The horizontal limits correspond to the low fuel rates, when the mixing is assumed to be completed within the recirculation region.

In a global sense, based on the dynamic pressure ratio of the fuel to the air at the thermodynamic conditions upstream of the flameholder, q_r, Fig. 6.24 shows a region of linear dependence of the equivalence ratio at blowout on the dynamic pressure ratio at high equivalence ratios. Otherwise, at low dynamic pressure ratios, the blowout limit appears to be insensitive to this parameter.

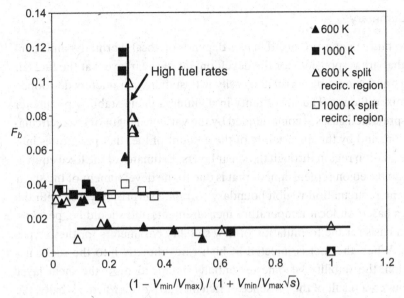

Figure 6.23. Blowout equivalence ratio vs. shear-layer growth parameter. V_{min} and V_{max} are the velocities across the shear layer, and s is the density ratio. The plateaus indicate mixing regimes that dictate the flame blowout. The vertical line corresponds to high fuel flow rates, indicating that large quantities of fuel leave the recirculation region before mixing is complete.

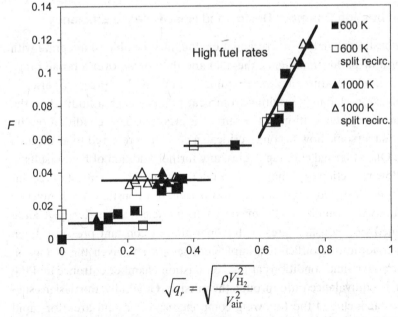

Figure 6.24. Blowout equivalence ratios appear insensitive to the jet-to-upstream air dynamic pressure ratio at low fuel rates and linearly dependent at large fuel flows.

6.5.6 Summary

Flame stability in non-premixed flames depends on local thermodynamic conditions that are responsible for the development of shear layers at the fuel–air boundary and cannot be correlated easily with stability parameters developed for premixed flames. The uncertainty in defining a flame stability parameter for non-premixed gases is compounded by the variable length of the recirculation region, and by the uncertainty of the amount of fuel that penetrates into the recirculation region through the shear layers. Estimates of the local equivalence ratio based on replenishment that is due to the development of the shear layer at the recirculation-region boundary correlate surprisingly well with the estimates based on local temperature measurements and should be, perhaps, used as a basis for the formulation of new stability parameters for these types of flows. When the fuel is injected at high rates directly into the recirculation region, the stability becomes essentially independent of the shear-layer development, a result of the presence of rich mixtures even when, globally, the fuel rates are low. Therefore, stability parameters, determined primarily from global data, fail in general to describe the flameholding process. However, limited local information, acquired in the recirculation region itself, provides correction factors that reproduce some of the physical processes, in certain cases, with satisfactory accuracy.

6.6 Combustion Chamber Design and Heat-Release Efficiency

The combustion chamber of a scramjet engine must be able to integrate with the other flow-path components, the inlet and the nozzle, over a broad range of pressure, temperature, and flow-regime conditions. The pressure raises in the combustion chamber resulting from heat release reach a limit when the maximum energy deposition in the burner is attained. This condition occurs when the supersonic flow becomes critical, a condition referred to as thermal choking. Once thermal choking begins, any further addition of heat results in a mass-flow reduction and may lead to inlet unstart. The combustion chamber could be designed, theoretically, to operate at a constant Mach number close to unity to capitalize on the maximum heat release; alternatively, it could be designed with constant area or for operation at constant pressure. Each of these conditions is difficult to implement separately, given the change in flow thermodynamic conditions at the combustion chamber entrance and the variation in equivalence ratio through the mission. Generally, the designs suggested so far included the following components: a constant area for rapid heat release followed by an expansion that allows additional heat addition after thermal choking has taken place at the end of the constant cross-section area and a further diverging section that may be considered the internal

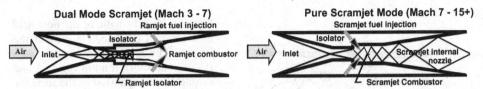

Figure 6.25. The combustion chamber design includes a constant area for rapid heat release followed by the expanding area, where additional heat can be released in ramjet operation. Further expansion takes place in the internal nozzle (schematic after McClinton, 2002).

nozzle leading to the external nozzle. An isolating section is needed between the inlet and the combustion chamber to accommodate the pressure differences between these two components. Often a step expansion is included at the combustion chamber entrance to offer additional separation while, at the same time, acting as a flameholding device and a thrust surface.

In the lower range of engine operation, usually below a flight Mach number of 7, the engine operation ranges from subsonic combustion and transitions to supersonic operation. Even when most of the combustion takes place in a supersonic flow, the combustion chamber includes regions of subsonic flow; large gradients of temperature and velocity are present. This operational condition is often referred to as a "dual-mode" combustion chamber. A shock train forms in the isolator as a result of the adjustment to the pressure rise in the combustion chamber, as shown in Fig. 6.25. A separation may appear that confines the shock train in a supersonic core that may extend into the combustion chamber. Because the flow residence time is relatively large in the low-Mach-number regime and combustion is expected to take place within a relatively short distance, the fuel is distributed in the expanding region of the combustion chamber. At a higher-Mach-number regime, the flow is supersonic throughout the combustion chamber. Fuel injection must begin at the entrance into the combustion chamber, and heat release takes place within the constant cross section until conditions close to thermal choking are encountered.

The effects of pressure rise in the combustor and the influence in the subsonic flow regions extend upstream into the isolator section and modify the structure of the shock system there. Given the importance of maintaining the flow started in the inlet under all flight conditions, the operability of the isolator is critical and deserves particular attention.

6.6.1 Isolator

The isolator is a duct that in general plays no role other than protecting the inlet flow from adverse back pressure. There are concepts in which fuel could be injected in this segment of the engine (Ortwerth, 2000; Owens et al., 2001a); however, most of the isolator designs are simple ducts. The isolator

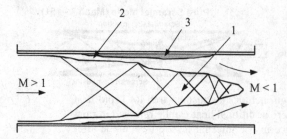

Figure 6.26. Schematic of the flow field in a constant cross-section area of the isolator. The flow is compressed through the oblique shock train in region 1 and through area expansion of the subsonic flow in region 2. A separation forms near the wall in region 3.

can be viewed as an extension of the inlet where additional compression takes place, but its purpose indicates a distinct function. In principle, the isolator adds weight, internal drag, and heat loads on the engine structure, and therefore its length must be limited to the minimum required by operability constraints.

In the complete absence of a boundary layer it can be possible to achieve the required pressure rise with a single, stable shock wave. But when a boundary layer is present, the interaction between the boundary layer and negative pressure gradient results in the boundary-layer separation and the formation of oblique shocks. The flow field in the isolator shown in Fig. 6.26 represents the case in which compression is achieved both through an oblique shock train in supersonic region 1 and through area expansion of the subsonic flow in region 2. Near the wall in region 3 is a separation that balances the pressure gradient across the isolator's length through the shear stress (Ortwerth, 2000). The separation appears when the first oblique shock creates a sufficient pressure rise to separate the boundary layer; a repeated shock structure follows in the core of the duct. The oblique-shock-wave train shown in Fig. 6.26 is characteristic for the higher-Mach-number operation. For moderate Mach numbers at the isolator's entrance, a weaker train of oblique shocks forms in the form of a set of lambda shocks with subsonic region 2 achieving a gradual compression in the isolator (Heiser and Pratt, 1994). In both cases the shock train must be such that the initial wave does not propagate upstream into the inlet, disrupting the flow or, in the limit, resulting in inlet unstart.

As a result of the formation of the shock-train structure, the issue of immediate interest is the prediction of the minimal isolator length that can be allowed. A second issue of great importance is the losses related to the isolator's drag. Both are subsequently reviewed briefly, following Ortwerth's (2000) analysis.

ISOLATOR LENGTH. The isolator's length is determined by the pressure rise that must be achieved between the inlet outlet and the combustion chamber entrance and, written in normalized parameters (Ortwerth, 2000), depends on the pressure ratio, $p_r = p_{out}/p_{in}$, as

$$\frac{X}{D_H} = \frac{1}{4K} \frac{g_1^2}{\gamma f_1} \left[\frac{p_r - 1}{(f_1 - p_r)(f_1 - 1)} + \frac{1}{f_1} \ln \frac{p_r(f_1 - 1)}{(f_1 - p_r)} \right] + \frac{\gamma - 1}{2\gamma} \ln p_r,$$

(6.45)

where $4K = \text{const} \times$ friction coefficient at the duct entrance; the constant is taken as 44.5 (Ortwerth, 2000); D_H is the hydraulic diameter, 4 area/duct perimeter; $f_1 = F_1/p_1 A_1$, where F is the stream thrust, A is the area, and the subscript refers to the isolator entrance station; and $g_1 = \dot{m}\sqrt{(\gamma - 1) H_0}/p_1 A_1$, where H_0 is the stagnation enthalpy at the duct entrance.

The accuracy of the shock-train length prediction by use of Eq. (6.45) was within 20% of a broad range of experimental results that included ducts of various shapes (round and rectangular), entrance Mach numbers ranging from 1.5 to 5, order-of-magnitude variation in the entrance Reynolds number, and different friction coefficients.

The correlation by Waltrup and Billig (1973), based on experimental data in a circular duct, resulted in the following equation:

$$\frac{X}{\sqrt{D}} = \frac{\theta^{1/2}}{\text{Re}_\theta^{1/4}} \frac{1}{M_1^2 - 1} [50(p_r - 1) + 170(p_r - 1)^2],$$

(6.46)

which emphasizes the shock-train length dependence on the Reynolds number Re_θ, the boundary-layer momentum thickness θ, and the isolator entrance Mach number M_1. Equation (6.46) indicates that, for a fixed pressure ratio, the shock-train length increases for flows with thick boundary layers at the isolator entrance and decreases with increased Reynolds and Mach numbers. Thicker boundary layers separate under weaker shock waves, and the smaller angle of the initial shock wave results in a long shock train. As the Reynolds number increases, the boundary layer can withstand stronger shocks and the angle of the initial shock that generates the separation is more abrupt, resulting in a shorter shock train. The Mach number effect is not immediately evident from the equation because it is coupled to the pressure ratio. As the Mach number increases, the required compression that would lead to separation is higher. The angle of the shock wave that leads to separation is lower, and the shock train is longer. These effects are well illustrated by the experiments described by Chinzei et al. (2000) and reproduced in Fig. 6.27. The difference in thrust based on pressure integrals relative to a reference case, ΔF, decreases for a short isolator as the amount of fuel burnt in the combustion chamber increases and the pressure rise in the chamber leads to a more severe separation in the isolator. A longer isolator experiences less separation, and the thrust increment is larger than that for the short isolator. With a longer isolator the thrust increases initially as more fuel leads to higher heat release, and the isolator can accommodate the pressure rise without severe penalties

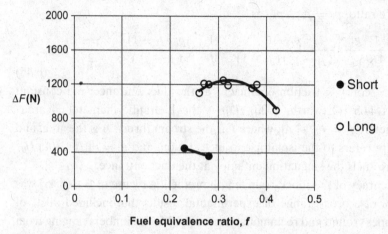

Fuel equivalence ratio, *f*

Figure 6.27. Effect of pressure rise through heat release in the combustion chamber on the thrust increment through the separation encountered in the isolator. A longer isolator can accommodate higher pressure rises with less separation, and thrust continues to increase as the fuel releases more heat in the combustion chamber. The negative effects of separation in the isolator are delayed to higher equivalence ratios.

on the boundary-layer separation. As the fuel flow rate continues to increase, the long isolator also begins to experience the negative effects of a large pressure rise and significant separation.

ISOLATOR DRAG LOSSES. The main sources of losses in the isolator are caused by the pressure drag and the viscous drag. In terms of entropy generation, these losses lead to

$$\frac{ds}{R} = \left(\frac{p_w}{p} - 1\right)\frac{dA}{A} + \frac{\tau_w d\sigma}{pA}, \tag{6.47}$$

where p_w is the wall pressure at the station selected for analysis, τ_w is the wall shear stress, A is the cross-section area, and $d\sigma$ is an element of duct wall area.

If a pressure drag is defined in terms of a drag coefficient, C_d, the drag results from

$$\frac{dD}{d\sigma} = \frac{C_d 1/2\rho V^2 A}{S_w}, \tag{6.48}$$

where S_w is the isolator's wetted area. Equation (6.48) then defines the isolator's drag.

The flow in the isolator as previously described refers to the case in which the entrance is both uniform and parallel to the axial direction. It also assumes that the boundary layer is the same on all sides, which, in general, will not be the case for a realistically shaped inlet ingesting part of the forebody's boundary layer. Any departure from symmetric entrance conditions may change the

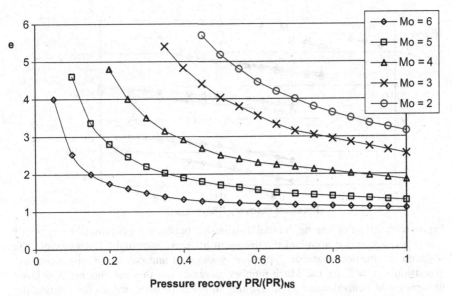

Figure 6.28. Required combustion chamber expansion with thermal choked conditions at the end of the constant-area duct as a function of the pressure recovery and the flight Mach number.

flow in the isolator by the formation of uneven separations on one side or another in the isolator. In this case the length of the shock train can be affected to a substantial degree.

6.6.2 Combustion Chamber Design and Performance

6.6.2.1 General Chamber Design Parameters

In the most general case, a supersonic combustion chamber includes an area expansion ε from the isolator cross-section area to a constant-area duct followed by an expanding section similar to the diagram shown in Fig. 6.25. For this generic configuration, Ortwerth (2000) gave an analysis that estimates the required area expansion from the isolator to the combustion chamber when operating at a stoichiometric mixture ratio and assuming that the flow remains supersonic but close to thermal choking toward the end of the constant-area section. The expansion delays the upstream interactions caused by thermal choking. Shown in Fig. 6.28, the expansion depends on the pressure recovery – with the baseline considered the recovery of the normal shock – and the flight Mach number. This excludes the presence of a sudden expansion often used to create a recirculation region to enhance flameholding.

Clearly the pressure-recovery improvement reduces the chamber expansion requirement; because the static pressure at the beginning of the heat-release section is lower when the pressure recovery is higher, there is less

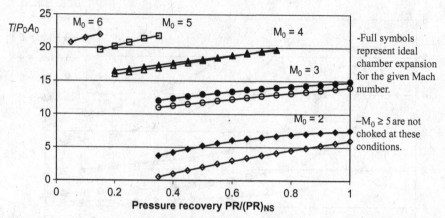

Figure 6.29. Effect of staging the fuel distribution between the combustion chamber's constant cross-section area and the divergent area. At each flight Mach number the required combustion chamber expansion is assumed and compared with fixing the expansion to $\varepsilon = 2$. As the Mach number increases the thermal choking is delayed to higher fuel equivalence ratios, resulting in fewer thrust penalties for limiting the chamber expansion. Beyond $M_0 = 5$ fuel distribution in the expanding section is not necessary.

need to protect the isolator's flow from the pressure rise in the combustion chamber.

The effect of the flight Mach number M_0 on the expansion size is considerable. Thermal choking appears more rapidly when air enters the combustion chamber with a lower stagnation temperature because the relative enthalpy increase through combustion is larger (Segal et al., 1995), and, as a result, the expansion must be larger at lower Mach numbers.

Overall, combustion chamber expansions larger than $\varepsilon = 2$ are impractical (Ortwerth, 2000) because they would result in an insufficient compression ratio for high flight Mach numbers.

In the divergent part of the combustion chamber, additional fuel can be injected, provided that the constant cross-section area has not reached thermal choking and the flow continues to remain supersonic. In fact, at low flight Mach numbers, when the constant cross-section area is susceptible to becoming critical with relatively modest rates of heat release, it is necessary to divert some, or all, of the fuel to the expanding section. Fortuitously, this situation is one in which the residence time in the combustion chamber is higher than during flight at high Mach numbers, and therefore the length required for complete combustion is shorter and fuel can be burned completely within the divergent section. As a figure of merit, the analysis of the specific thrust (Ortwerth, 2000) $T/P_0 A_0$, where P_0 is the flight altitude pressure and A_0 is the isolator entrance area, is shown in Fig. 6.29 as a function of the pressure recovery normalized to that of the normal shock. This analysis considers the case in which the combustion chamber expansion is limited to $\varepsilon = 2$. The results are compared with the case in which the ideal expansion, based on the solutions in Fig. 6.29, could be tolerated at each of the flight Mach numbers considered in the analysis. It is evident that, below $M_0 = 4$, limiting the combustion chamber expansion and therefore having to distribute the fuel between the constant area and the divergent area results in a relative reduction of specific thrust. As the flight Mach number increases, less fuel has to be distributed in the divergent section and limiting the combustion chamber expansion has a weaker effect on the specific thrust. In fact, at the conditions of this particular analysis, the constant cross-section area does not become thermally choked at $M_0 \geq 5$.

This simplified analysis applies to a generalized design concept that includes the three elements, previously described. The domain of design configurations is certainly far broader and depends on the vehicle's anticipated mission and the designer experience. The following subsection presents a small sample of supersonic combustion chambers' results obtained in various studies. The space available here clearly cannot include a detailed list, and many interesting and insightful studies are left out.

6.6.2.2 Pressure Rise and Combustion Efficiency

A measure of efficiency is the degree of energy conversion into heat within the combustion chamber, and it is well captured by the chamber pressure rise. Other parameters that affect the pressure distribution include wall heat transfer, the presence of shock waves, wall separations, etc. Because wall pressure is both a reliable and a convenient measurement in an environment that does not easily tolerate intrusive measurements, it is almost always a reference measurement in supersonic combustion applications.

The studies of supersonic or dual-mode combustion are numerous. They include both experimental and computational results in ground facilities and data acquired during several flight tests. The discussion included in the next subsection uses examples from only a few of these studies selected to outline combustion chamber design requirements. For a more detailed list of experimental results and the measured combustion chamber performances, a good starting point is the review by Murthy (2000).

CONSTANT AND DIVERGING CROSS-SECTION COMBUSTION CHAMBERS. The constant cross-section area combustion chamber is the simplest geometrical configuration and offers a limiting case study because the heat release in a constant area leads to a rapid pressure rise; the early onset of thermal choking limits the allowable energy-deposition amount before upstream interaction occurs. In the experiments described by Falempin (2000), hydrogen wall injection in a Mach 2.8 flow at stagnation temperatures in excess of 2000 K result in a rapid and continuously increasing wall pressure, even for low equivalence ratios (below 0.5). This pressure gradient is difficult to control, and any further slight increase in equivalence ratio, or a reduction in the air stagnation temperature, can result in upstream interaction and inlet unstart.

The addition of a divergent section delays the pressure rise and could theoretically provide a region of constant-wall-pressure distribution as the area expansion compensates for the heat-release effect on the flow Mach number (Heiser and Pratt, 1994). As the rate of heat release diminishes farther downstream, the continuously diverging area results in a continuous drop in the pressure distribution. In practice, the constant-pressure region is difficult to materialize, as was indicated by several experimental studies (as examples, see Northam et al., 1992; Billig, 1993; Falempin, 2000).

The simplified analysis by Billig (1988) assumed a control volume that included the combustion chamber from entrance to exit that allows for conditions when the flow is

1. supersonic throughout the chamber,
2. supersonic at the entrance and a normal shock is present within the control volume, or
3. supersonic flow throughout the control volume with a possible wall separation that is entirely contained within the control volume.

Fuel is injected in this control volume, and the chemical reactions are assumed to reach equilibrium; the wall shear and heat transfer can be estimated from the conventional relationships for the friction coefficient C_f and the Stanton number $\mathrm{St} = q_w/\rho u h$, which relates the wall heat flux to the inviscid energy flux.

Under these assumptions the applicable conservation equations are

$$\rho_a u_a A_a + \dot{m}_f = \rho_b u_b A_b,$$

$$p_a A_a - p_b A_b + \int_a^b (p_w \sin\alpha - \tau_w \cos\alpha)\, dA_w = \rho_b u_b^2 A_b \tag{6.49}$$

$$- \rho_a u_a^2 A_a - \rho_f u_f^2 A_f \cos\beta,$$

$$h_a + u_a^2/2 + f(h_f + u_f^2/2) = (1+f)(h_b + u_b^2/2) + q_w A_w/\dot{m}_a,$$

where the subscripts a, b, f, and w refer to the control-volume entrance, exit, fuel, and wall, respectively; α and β are the wall and the fuel-injection angles, respectively, with the axial direction. With the addition of the equation of state, the system is complete and offers solutions for the wall pressure distribution.

The pressure-area distribution is taken in the form

$$P A^{\varepsilon/\varepsilon} - 1 = \text{const}, \tag{6.50}$$

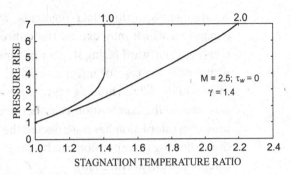

Figure 6.30. Pressure rise as a function of heat release and area divergence (after Billig, 1988).

which is applicable in the downstream section of the combustion chamber where the assumption of one-dimensionality can be reasonably made. Here, ε is an exponent that depends on the degree of heat release in the combustion chamber and that assumes values of one for a constant-area device, zero for a constant-pressure operation, and $-\gamma M^2$ for a constant Mach number (Billig, 1988).

In the particular case in which the heat release has ceased to develop, $dT_{\text{stag}}/T_{\text{stag}} \to 0$, and, additionally, the flow can be assumed isentropic with negligible wall shear, a simple expression can be found for ε in the form

$$\varepsilon = \frac{\gamma M^2}{1 + (\gamma - 1)M^2}. \tag{6.51}$$

The cross-section area effect is clearly seen in Billig's analysis (1988), which resulted in the curves reproduced in Fig. 6.30. This analysis, which is based on zero wall shear, adiabatic walls, and constant γ for a combustion chamber entrance $M = 2.5$, indicates that the pressure rise for a stagnation-temperature ratio increase from 1.2 to 1.4, for example, with an area ratio expansion of 2, the pressure rise experienced is approximately 44% larger, whereas if the area cross section is maintained constant the pressure rise is 83% larger. More significant, a further increase of heat release causes an asymptotic pressure rise in the case of the constant-area combustor, which would result in upstream interaction.

The usefulness of this calculation derives from the ability to calculate the pressure rise for the given heat release shown in Fig. 6.30, and hence it allows designing the isolator shock train accordingly to match the expected pressure rise through combustion. In a sense, this calculation is similar to a Rayleigh flow analysis – which was discussed in Chap. 2 – with the observation that some solutions may result in a pressure rise larger than the normal shock assumption (Billig, 1988). In this case the slope relationship is no longer enforced, and the calculation is based on the subsonic solution.

COMBUSTION CHAMBERS WITH SUDDEN AREA EXPANSION. Most practical solutions will have a combustion chamber that includes sudden area expansions, constant-cross-section duct segments, and divergent sections. The sudden expansion, which may extend the entire width of the combustion chamber, creating a rearward facing step, or be limited to only a portion of the chamber width, as is the case of injection ramps, creates a base flow that, in addition to providing a flameholding region, provides a separation between the pressure rise in the combustion chamber and the isolator's flow. As long as the upstream interaction has not affected the isolator's flow, the pressure distribution following an expansion can be generically included in the following three categories (Ortwerth, 2000):

- The low-flight-Mach-number regimes: These are characterized by relatively low enthalpy at the combustion chamber entrance and are conducive to a rapid pressure rise with modest amounts of heat addition. Thermal choking effects are dominant, causing the pressure to rise rapidly behind the sudden expansion and to remain almost constant until the exit.
- The intermediate-enthalpy flows in which the pressure rises behind the sudden expansion above the isolator's pressure and its axial distribution is dictated by the efficiency of the combustion process.
- The high-enthalpy case in which the relative amount of heat release to the airflow enthalpy is smaller than in the preceding cases and the sudden expansion may actually lead to a pressure drop from the isolator's level. In this case, the flow experiences a pressure rise at the reattachment point and the base is maintained at pressures that are lower than the reattachment pressure.

A configuration that included several of the elements just described was used in the experiments by Owens et al. (2001b). This configuration included a constant-cross-section isolator with a 6:1 length-to-height ratio followed by a combination of ramp and sudden expansion with fuel injected from the ramp and the expansion in an axial direction, followed by a 6° half-angle diverging duct. The air Mach number at the isolator's entrance was 1.6, and the stagnation temperature was 850 K. The expansion was $\varepsilon = 2$ and the constant-cross-section segment length was short, extending over only five isolator duct heights. As a result, a significant amount of heat was released in the diverging duct, and upstream interactions were not encountered even when the equivalence ratio was increased above 0.6. However, a penalty was paid in terms of the combustion efficiency estimated at the end of the constant-cross-section duct. Figure 6.31 shows the pressure distribution for several combinations of equivalence ratios, with hydrogen supplied from both injectors. The plot shows the higher pressure rise obtained when fuel injection from the base was preponderent. The static pressure distribution had a maximum value near the center of the constant-area section and was associated with the reattachment location of the flow over the step. Downstream, the pressure dropped quickly

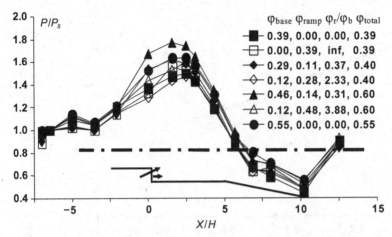

Figure 6.31. Streamwise pressure distribution profile (normalized) for φ_{total} up to 0.6, showing the absence of thermal choking. The axial direction is normalized by the isolator's height. The air Mach number at the isolator's entrance is 1.6 and the stagnation temperature is 850 K. Several pressure measurement locations are shown in the isolator duct.

and continuously through the constant-area combustor and into the expansion section until near the end, where separation shocks occured to match the local ambient pressure.

Acceleration of the air in both the second half of the constant-area duct, i.e., beyond $2.5H$, and the expansion duct indicated supersonic expansion section flow. The presence of low-speed, subsonic-burning layers in the vicinity of the walls (Strokin and Grachov, 1997) generated a convergent–divergent channel in which the core flow remained supersonic throughout the constant area of the test section and then continued to expand in the divergent section. As a result, the favorable pressure gradient tended to decrease the shear-layer growth in the fully supersonic case of these experiments, and upstream shock interactions were not present.

The combustion efficiency was defined as $\eta_c = \varphi_r / \varphi_t$, where φ_r is the reacting equivalence ratio and φ_t is the total injected equivalence ratio. Here, the reacting equivalence ratio was calculated with a 1D analysis that determined the amount of fuel entirely consumed to raise the pressure to the value measured in the experiment at each axial location, whereas the rest of the fuel was considered mixed in the air under the local thermodynamic conditions. Figure 6.32 shows the combustion efficiency at the end of the constant-area duct. As the efficiency dropped with increased total equivalence ratio, operation with $\varphi = 0.7$ resulted in stable and controllable combustion without thermal choking. The analysis showed a definite transition from a relatively high level of performance up to an equivalence ratio of 0.1 to a lower level at $\varphi = 0.5$. The estimated combustion efficiency, which incorporated both chemical kinetics and mixing effects, was about 60% at $\varphi = 0.5$, further dropping as

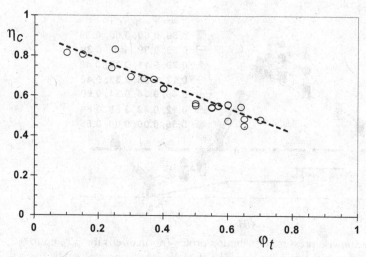

Figure 6.32. Combustion efficiency based on wall pressures.

the total equivalence ratio increased. This reduction in combustion efficiency was attributed to reduced mixing and is not expected for normally propagating flames for which flame spreading should increase as heating decelerates the entering air.

Combinations of the combustion chamber elements, including ramps, sudden expansions, and divergent sections, are shown in Fig. 6.33 as evaluated by Northam et al. (1992) in a study that connected the combustion chamber directly to the facility's supersonic nozzle. The fuel was injected in an axial direction from the ramps; this is an advantageous injection solution because it alleviates the losses associated with shock-wave formation during transverse injection. At the same time, this type of fuel injection removes the fuel from the vicinity of the walls, and it creates vortical motion that enhances mixing. Additionally, cycle analyses indicate that, when the flight Mach number exceeds a value of approximately 10, the fuel axial momentum becomes a significant contribution to the overall thrust (Billig, 1993). This type of injection is often referred to as parallel injection – to differentiate it from the transfer injection – although the fuel enters the combustion chamber at an angle. Shown in Fig. 6.33 are three selected configurations evaluated during this study, which consisted of a constant-cross-section area isolator that contained the ramp injector followed by a constant-cross-section duct and a sudden expansion along with other arrangements that included combinations of a constant-cross-section area with divergent sections of different lengths and expansion angles. The fuel-injection location was moved downstream in the configurations and did not benefit from the sudden expansion, as is the case in configuration 1.

The wall pressure distribution is shown in Fig. 6.34 for configuration 2, normalized by the static pressure at the isolator entrance, for hydrogen–air

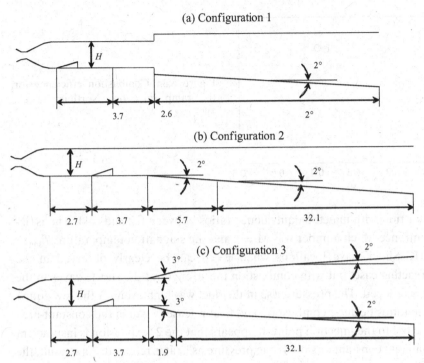

Figure 6.33. Based on evaluation from Northam et al. (1992). The isolator entrance $M = 2$, the stagnation temperature $T_0 = 1700\,\mathrm{K}$, and the fuel was injected from the swept ramps with the equivalence ratio indicated in the figure. Shock waves are evident when $\varphi = 0$ but diminish with combustion. The high pressure rise in the duct, for the relatively low equivalence ratio, is due to the early heat release beginning in the constant-area duct. Upstream interaction appears as soon as the equivalence ratio increases above 0.4 under these conditions and propagates into the isolator with only a small further fuel flow-rate increase.

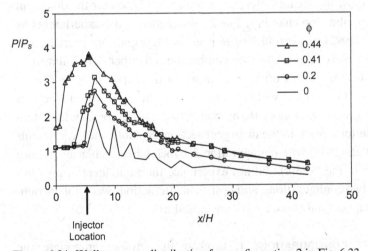

Figure 6.34. Wall pressure distribution for configuration 2 in Fig. 6.33.

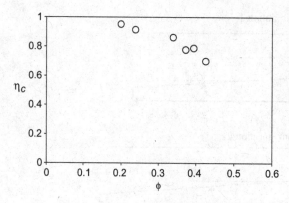

Figure 6.35. Combustion efficiency for the ramp injection (Northam et al., 1992).

combustion with injected equivalence ratios between 0.2 and 0.44. The isolator entrance Mach number was M = 2 and the stagnation temperature $T_{stag} = 1700$ K. Shock-wave locations when $\varphi = 0$ can be clearly observed in the nonreacting case, but with combustion the strength of the shock waves in the duct is reduced. The pressure rise in the duct was high even for the low equivalence ratios because combustion and heat release began in the constant-area duct close to the injection point. It appears that the 2° expansion is insufficient to maintain constant or slow-rising pressure as heat is released. As a result, the upstream interaction appears as soon as the equivalence ratio increases above $\varphi = 0.4$ under these operational conditions and it propagates upstream of the isolator with only a small further fuel-flow-rate increase.

The combustion efficiency estimated for these design solutions was high, as shown in Fig. 6.35, as expected for the high pressure rise measured in the combustion chamber.

Despite the obvious differences in flow Mach number and stagnation temperatures at the isolator entrance between the two sets of experiments just described, the most significant difference is the angle chosen for the divergent section of the combustion chamber. The 2° divergence in the experiments by Northam et al. (1992) was insufficient to prevent the significant pressure rise that was due to heat release and the combustion chamber acted almost as a long constant-cross-section duct. In contrast, in the experiment by Owens et al. (2001b), the heat release within the short constant-cross-section duct was insufficient to bring the flow close to thermal-choking conditions, and substantial heat release took place in the divergent section. Notably, the experiments by Northam et al. (1992) with configuration 3, which had a higher divergence angle as shown in Fig. 6.33(c), did not experience the same level of pressure rise as the first two configurations, and upstream interactions were not encountered even when the equivalence ratio was raised to $\varphi = 1.2$.

MULTIPLE-STRUT CONFIGURATIONS. It is clear that, to accommodate the requirement for efficient heat release in the combustion chamber throughout

Figure 6.36. The CDE prepared for testing in NASA's 8-ft. high temperature tunnel.

the entire range of flight conditions, at both high speeds and low speeds, the fuel injection must be distributed along the constant area and the diverging section of the combustion chamber. Furthermore, limitations of fuel penetration and mixing determined by the air–fuel interactions, described in the mixing discussion in Section 6.4, indicate that a relatively small duct height is advantageous and that this can be achieved for an engine with a required capture area by dividing the engine duct into segments, each with separate airflow path and fuel-injection distribution. Such a configuration was used in NASA's concept demonstration engine (CDE), shown in Fig. 6.36, which was evaluated as part of the NASP Program (Kumar et al., 2001).

The strut design can be such that a 3D oblique shock system is formed at the strut's leading edges by careful selection of the angle to achieve both a high inlet pressure recovery and a high area-capture ratio. At the same time, the fuel injectors can be distributed in both the axial and the transverse directions to the strut height, as was suggested by Ferri (1973), to enable fuel flow modulation throughout the entire operational range without having to resort to area changes. One or more adjacent struts attached to a common inlet and common nozzle would form an engine module.

An example of an internal duct geometry that resulted from the use of struts is shown in Fig. 6.37, as was used by Goldfeld et al. (2001) in free-jet tests for a range of Mach numbers from 2 to 6. Here, in addition to the axial and the transverse distributions of the fuel injectors, small ramps penetrating into the flow were used to improve fuel penetration and mixing. High-pressure recovery was achieved in this configuration, and the capture ratio A_0/A_c was both high and uniform for free-stream Mach numbers ranging from 3 to 6.

A series of experiments on a strut-configured engine module by Kanda et al. (1997) in the Mach 6–8 range indicated that stable operation can be achieved over a broad range of experimental conditions with high efficiency,

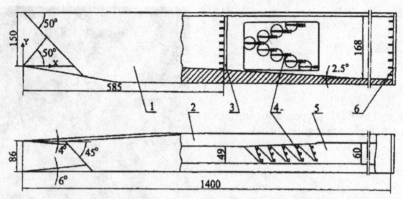

Figure 6.37. Scramjet module evaluated by Goldfeld et al. (2001): 1, engine side wall; 2, internal strut from which fuel is injected; 3, stagnation pressure rake; 4, fuel-injector ramps; 5, the constant cross section of the combustion chamber; and 6, exit stagnation pressure rake. All dimensions are in millimeters.

reaching 90% for conditions around the stoichiometric fuel–air ratio. Yet the thrust level was modest, and inlet unstart was experienced in certain cases in which the fuel flow rates were high or the Mach number low, pointing to the still substantial difficulties in designing a successful scramjet engine that satisfies both efficiency and performance requirements throughout the entire flight envelope.

6.7 Scaling Factors

Much of the scramjet developmental work is based on measurements made with scaled models in ground facilities because the technical difficulties and the costs associated with flight testing limit the ability to acquire sufficient in-flight data. Despite great advances in theoretical modeling of the physical processes, detailed computational analyses are also demanding, and most of the models currently used lack sufficient experimental validation. As a result, the test objects are, in general, smaller than the anticipated flying devices, and the thermodynamic conditions under which testing is done in ground facilities cannot always reproduce real flight conditions. For these reasons, scaling laws are needed to account for both (a) geometry sizing and (b) changes of the thermodynamic conditions under which the engine operates. These scaling laws are expected to indicate the operational conditions that are required for testing on reduced-scale models to produce the same performance as the full-scale scramjet in flight.

For a constant-geometry scramjet engine, the parameters that dictate the thermodynamic state of the flow and, implicitly, the performance measured either by specific impulse or combustion efficiency are pressure and temperature profiles, the wall heat transfer, velocity distribution, and gas composition.

A first-order simulation requires reproducing the Mach number M, the Reynolds number Re, the Stanton number St, and the Damköhler numbers Da_1 and Da_2 (Anderson et al., 2000). All these parameters have been mentioned on various occasions before but are reproduced here to identify the physical parameters that will have to appear in the scaling process:

$$M = \frac{\text{velocity}}{\text{speed of sound}} = \frac{u}{\sqrt{\gamma RT}},$$

$$Re = \frac{\text{inertial forces}}{\text{viscous forces}} = \frac{uL}{\upsilon},$$

$$St = \frac{\text{wall heat flux}}{\text{core flow energy flux}} = \frac{q_w}{\dot{m}H}, \qquad (6.52)$$

$$Da_1 = \frac{\text{flow residence time}}{\text{chemical-reaction time}} = \frac{kL}{u},$$

$$Da_2 = \frac{\text{diffusion time}}{\text{chemical-reaction time}} = \frac{kL^2}{D_{ik}}.$$

Additional parameters play a role, such as the turbulent fluctuations described in a nondimensional form by the third Damköhler number, $Da_3 = \sqrt{\varepsilon/\mu Da_1}$, where ε is the turbulent eddy viscosity and μ is the laminar viscosity (Inger, 2001).

Not all of the physical parameters that appear in the similarity parameters listed in Eqs. (6.52) can be satisfied simultaneously because model size, velocity, and temperature cannot be changed in a fashion that would maintain the nondimensional scaling numbers. Because the model size must be scaled because of the physical constraints of existing facilities and the gas composition must be maintained to reproduce the basic mixing and combustion interactions, the other parameters, namely temperature and pressure, must be adapted in scaled model testing.

The study by Pulsonetti and Stalker (1996), using two models with a length ratio of 5, offered revealing scaling rules for the mixing and combustion processes. When the mixing of parallel fuel and airstreams through the development of turbulent compressible shear layers is of interest, a characteristic length can be considered the distance required to achieve a mixing at the molecular level. That distance was found by Pulsonetti and Stalker (1996) to scale inverse proportionally with the pressure. Similar pressure–length dependence was found when considering the wall friction and heat transfer that are determined by the boundary-layer development that, in turn, is largely dependent on the Reynolds number. Therefore a scaling rule that maintains the pressure–distance constant was recommended by the analysis of Pulsonetti and Stalker (1996):

$$PL = \text{const.} \qquad (6.53)$$

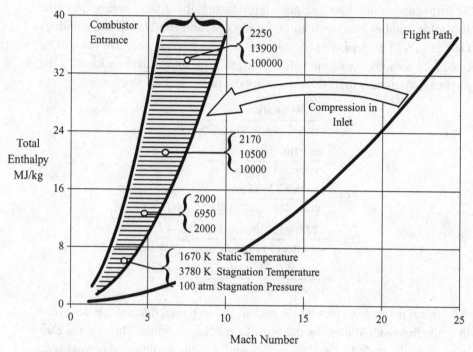

Figure 6.38. Total enthalpy, pressure, and temperature requirements for a combustion chamber simulation.

If a simplified hydrogen–air ignition model is used, as described in Section 6.4, ignition time can be considered to be dominated by radical formation reactions, and therefore ignition time would vary inverse proportionally with pressure and thus the pressure–length scaling would still be valid. The situation is different, however, when considering the combustion progress toward formation of stable products because the reaction rates depend on the pressure. In this case, (pressure)[power coefficient] – length would be a more appropriate scaling. Depending on the flight regime, specifically on the temperature and the flow velocity, these reactions can be fast or slow when compared with the residence time, and therefore the selection of the scaling rule for experimental evaluation on scaled models in ground facilities must be made according to the flight regime under simulation.

It should be noted that because the mass flow is determined by $\dot{m} = \sqrt{\gamma/RT}\,pA$M, if the Mach number and the temperature are kept the same in the simulation, the mass flow will scale linearly proportional with the model length scale.

Whether the desired thermodynamic parameters scaling can be accomplished depends on the ability to reproduce flight conditions in ground facilities. Figure 6.38 shows a calculation by Anderson et al. (2000) in which a flight path for a scramjet-equipped vehicle was selected and a reasonable compression in the vehicle inlet was assumed to decrease the Mach number at the combustion chamber entrance. The range of static temperature, stagnation temperature, and stagnation pressure are shown at several selected flight conditions. These conditions must be reproduced for an accurate flight-condition simulation at the combustion chamber entrance. It can be seen in this analysis that even for moderate hypersonic flight conditions the total enthalpy and stagnation pressure are very high, and few facilities can provide these experimental conditions. Shock tubes and expansion tunnels that can reproduce this level of flight enthalpy have a short time duration, limited to milliseconds or even less and therefore cannot accurately reproduce many of the relevant processes, including wall heat transfer. For that reason appropriate scaling must be used in ground facilities on subscaled models with correctly adjusted experimental conditions.

6.8 Fuel Management

6.8.1 Fuels as Vehicle and Engine Component Coolant Agents

Broadly classified, hydrogen and hydrocarbon fuels constitute two distinct groups of fuels that find applications for different regions of the scramjet operational regimes. Given the high temperatures experienced by the hypersonic vehicle's leading edges and several of the engine components, active cooling will be required. This function must be taken by the fuel flow because an additional cooling agent on board and the associated heat exchangers would be prohibitively heavy. Besides, if part of the aerodynamic heating is recovered by the fuel and later deposited in the engine, it may be argued that using the fuel as a cooling agent contributes to reducing the vehicle drag. The question is then whether the fuel flow required by the engine to produce thrust is sufficient to provide the heat sink required by the vehicle components. If the answer is negative, it would imply that the fuel flow used as a cooling agent must exceed the level required for thrust, and the engine would then operate wastefully with an excessively rich mixture. These competitive requirements define, in large part, the fuel selection for a specific mission.

Hydrogen has large heat-sink capabilities and, along with its fast chemical kinetics, makes a good candidate for operation at the high range of the hypersonic regime. Hydrocarbons have higher densities, which would lead to a smaller and lighter vehicle, and are easier to operate and store, but their heat-sink capability is limited; therefore these fuels are suitable for the lower range of hypersonic flight (Medwick et al., 1999; Bouchez et al., 2002).

Figure 6.39 reproduces the analysis by Heiser and Pratt (1994), which estimated the cooling capacity required for a selected scramjet engine as the

Table 6.4. *Selected fuel properties (CRC, 1988; Lander and Nixon, 1971)*

Fuel	Formula	Flash point (°C)	Freeze point (°C)	Density (kg/m^3)	Heating value (MJ/kg)	Heat sink at 1000 K (MJ/kg)
Hydrogen	H_2	Gas	−259	74.7 (liquid)	117.8	15.1
JP-7	$C_{12}H_{24}$ (avg)	69	−44	780	43.8	2.7*
MCH	$C_6H_{11}CH_3$	−3	−126	761	43.2	4.56*

* Includes both physical and chemical heat-sink capabilities.

flight Mach number increases. This analysis does not include the cooling load required by the airframe components. Initially, at low-Mach-number flight, the fuel flow needed for thrust generation exceeds the heat-sink demand, but as the flight Mach number increases, the heat loads increase rapidly, and the fuel flow required for thrust generation soon becomes insufficient to provide an adequate heat sink. For hydrogen, this limit is around Mach 15, whereas for JP-7 or methylcyclohexane (MCH) this limit is reached around Mach 10. When the airframe cooling load is considered, these limits are further reduced. The hydrogen higher heat-sink capability, $c_p \Delta T$, is due to higher constant-pressure specific heat, which is approximately six times larger than that of hydrocarbon fuels. This difference of heat capacity is somewhat reduced when the chemical heat-sink capability of *endothermic* fuels is considered. Endothermic fuels are those that absorb heat while undergoing a chemical decomposition into products that are themselves fuels, thus enhancing the fuel-cooling capability beyond the physical heat capacity that corresponds solely to its sensible enthalpy. Several relevant properties of hydrogen and selected endothermic hydrocarbon-fuel candidates are listed in Table 6.4.

The decomposition of the endothermic fuels begins at temperatures of around 800 K and the chemical heat sink rapidly increases, becoming, in

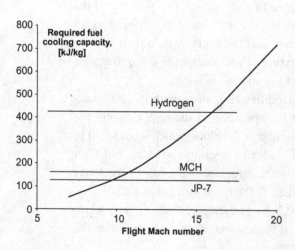

Figure 6.39. Engine components' heat-sink requirement for a typical scramjet flight trajectory with a constant flight dynamic pressure, $q_0 \cong 10^5$ N/m^2 (Heiser and Pratt, 1994).

Table 6.5. *Endothermic fuel reaction types and the corresponding heat sink (Maurice et al., 2000)*

Endothermic fuel decomposition	Reaction type	Theoretical chemical heat sink (kJ/kg)	Calculated heat of combustion of the endothermic reaction products (kJ/kg)
$C_6H_{11}CH_3$ (MCH) $\rightarrow C_7H_8$ (toluene) $+ 3H_2$	Dehydrogenation	2190	45 800
C_7H_{16} (*n*-heptane) $\rightarrow C_7H_8$ (toluene) $+ 4H_2$	Dehydrocyclization	2350	47 300
$C_{10}H_{12}$ (dicyclopentadiene) $\rightarrow 2$ c-C_5H_5	Dedimerization	621	43 630
$C_{12}H_{24}$ (kerosene) $\rightarrow CH_4$, C_2H_4, C_2H_6, etc.	Cracking	<3560	47 200
$CH_3OH \rightarrow CO + 2H_2$	Dehydrogenation	4000	20 420
$2CH_4 \rightarrow C_2H_2 + 3H_2$	Addition-dehydrogenation	11 765	62 860
$C_6H_6 \rightarrow 3C_2H_2$	Aromatic ring fracture	7650	48 280
$C_{10}H_{18}$ (decalin) $\rightarrow C_{10}H_8$ (naphthalene) $+ 5H_2$	Dehydrogenation	2210	40 700

certain cases, comparable with the physical heat sink, thereby essentially doubling the fuel heat-sink capability (Ianovski et al., 1997).

The amount of heat absorbed by the chemical decomposition is 26% to 31% of the total heat-sink capacity listed in Table 6.4 for JP-7 (Lander and Nixon, 1971; Huang et al., 2002) and 48% for MCH (Lander and Nixon, 1971). Other hydrocarbon compounds also exhibit large percentages of heat sink that are due to endothermic reactions, with JP-8+100 measured at 28% and JP-10 at 21% (Huang et al., 2002).

There is thus a clear advantage offered by endothermic fuels, provided that the chemical decomposition is not accompanied by the formation of undesired deposits in the cooling passages that reduce the heat transfer efficiency and may block the flow area; in this regard, hydrogen offers the added advantage of thermal stability. The key for the development of an efficient cooling system based on endothermic reactions is held by the paths of fuel decomposition and the compatibility with the materials that make the heat exchanger. Both issues are subsequently discussed.

A list of selected endothermic decomposition reactions is given in Table 6.5 (after Maurice et al., 2000) and includes the values of both the chemical heat sink and the heat of combustion of the products resulting from the endothermic reaction; the ratio of the heat sink to the heat of combustion is a true measure of the fuel-cooling capacity. In that regard, Ianovski et al. (1997) have shown that certain endothermic fuels reach 72% of the cooling capacity of cryogenic hydrogen when the reactor – namely the cooling channels in the vehicle or engine components – is maintained at elevated temperatures, around 1000 K. Overall, taken together, the physical and the chemical heat sinks of endothermic fuels reach close to 10% of their heat of combustion (Ianovski et al., 1997; Maurice et al., 2000). The means to achieve the fuel decomposition may be through thermal cracking or through catalytic decomposition facilitated by the reactor's material.

6.8.2 Thermal versus Catalytic Decomposition

The endothermic decomposition can be accomplished through thermal reactions or can be facilitated by the presence of an appropriate catalyst. Despite the increased mechanical complexity of the system, the catalytic decomposition offers the ability to select, to a certain degree, the products of decomposition, and it lowers the temperature at which the system operates (Lander and Nixon, 1971). Moreover, some thermal reactions that could potentially provide a substantial heat sink do not proceed along the expected paths and are accompanied by additional reactions of an exothermic nature that are evidently undesirable when a heat-sink role is demanded from the fuel because they reduce the cooling process efficiency (Lander and Nixon, 1971). For example, the decomposition of propane could take several different routes with different levels of heat-sink capability – even becoming exothermic – as the proportion of methane increases:

$$C_3H_8 \rightarrow C_3H_6 + H_2 \quad \text{absorbs } 2830 \, kJ/kg \, \text{(endothermic)},$$
$$C_3H_8 \rightarrow C_2H_4 + CH_4 \quad \text{absorbs } 1787 \, kJ/kg \, \text{(endothermic)},$$
$$2C_3H_8 \rightarrow 4CH_4 + 2C \quad \text{releases } 1016 \, kJ/kg \, \text{(exothermic)}.$$

It becomes clear that the application of endothermic fuels to a cooling engine and vehicle components would benefit from increasing the ability to select the reaction routes taken during the catalytic decomposition more than from enhancement of any particular reaction rate by increasing the thermal loading.

When the catalytic decomposition is selected, it is advantageous to select a reaction type that leads to the formation of an aromatic product that is stable because of resonance and is therefore thermodynamically preferred. Some *dehydrogenation* reactions that lead to the formation of aromatic compounds are shown in the examples listed in Table 6.5: MCH decomposes into toluene and hydrogen; decalin (and, similarly, JP-10) decomposes into naphthalene and hydrogen. *Cracking* of compounds, such as kerosene, may result in the formation of alkanes and alkenes, as indicated in Table 6.5, deriving from both endothermic and exothermic reactions that may reduce the conversion's heat-sink capability, as previously indicated.

If the initial hydrocarbon fuel is a system with sufficient carbon atoms, it may conceivably undergo a transformation to an aromatic compound

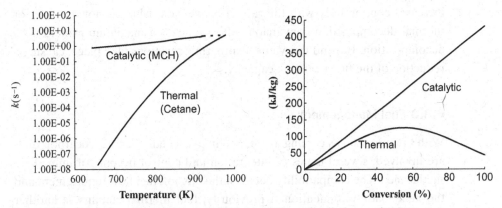

Figure 6.40. Catalytic vs. thermal decomposition reactions: Catalytic reaction rates are higher at low temperatures; the heat-sink capabilities of the catalytic reactions increase continuously with the rate of conversion whereas the thermal decomposition conversion reaches a maximum before exothermic reactions cause a reduction of the fuel heat-sink capacity.

with hydrogen release through a *dehydrocyclization* reaction. The *n*-heptane decomposition into toluene and hydrogen, which is listed in Table 6.5, is an example of this type of reaction. Aromatic compounds can form even from systems with low carbon content (Dagault and Cathonnet, 1998) through hydrogen abstraction followed by radical recombination. The formation of aromatic compounds is not without risk, because these compounds can lead to the formation of soot during combustion in the engine that, in turn, reduces the efficiency and may be deposited on the combustion chamber liner, increasing the radiative heat loads.

Catalytic reactions are more efficient than thermal reactions for an additional reason: Their reaction rates at low temperatures are greater (Lander and Nixon, 1971), thus extending the operational regime at which catalytic conversions can take place. This is an important advantage because the temperature at which the catalytic reactions progress is the temperature of the catalyst surface, not the fuel temperature, as is the case for thermal reactions. Extending the range of the regime for endothermic reactions toward lower temperatures is not the only operational advantage: The life of the catalyst is extended if the reactions can take place at lower temperatures.

The advantages exhibited by catalytic reactions over thermal decomposition are illustrated in Fig. 6.40 (from Lander and Nixon, 1971). The left part of the diagram indicates that catalytic reactions have high reaction rates over a broad range of temperatures. The thermal reactions proceed slowly at low temperature because of high activation energies required and only around 1000 K do their rates become comparable with those of the catalytic reactions.

The right-hand side of the diagram indicates the heat-sink capacity as a function of the degree of fuel decomposition. The catalytic reactions' heat sink increases continuously with the rate of conversion, which is not the case for thermal decomposition conversion, which reaches a maximum around 60% decomposition. Beyond this value, exothermic reactions take place, causing a reduction of the fuel heat-sink capacity.

6.8.3 Fuel Management

With the fuel used as cooling agent, and in particular when hydrocarbon fuels are involved, several issues of integration and control become of immediate significance. The compatibility between the mission fuel flow requirement and the heat loads was mentioned previously; fuel thermal stability is another major consideration. Reusable hypersonic vehicles are expected to operate over a long lifetime, and solid deposits on the fuel lines, heat exchangers, and engine components, which may result from hydrocarbon decomposition (Sobel and Spadaccini, 1997), may be a major technological difficulty. For missile applications, the accumulation of deposits in the cooling parts may not be an issue.

Formation of *coke*, the solid deposits resulting from the fuel chemical decomposition, is more severe as the heat exchanger operates at higher temperatures, as expected in hypersonic applications. At elevated temperatures, above 800 K, coke formation results from pyrolytic decomposition (Sobel and Spadaccini, 1997; Maurice et al., 2000; Huang et al., 2002), followed by recombination, leading to the formation of heavier hydrocarbons that then condense on the reactor's walls. Coke formation is expected to increase as the fuel energy density, deriving from the ratio of carbon to hydrogen atoms, increases. The appropriate use of a catalyst is thus critical. Platinum/alumina oxide is an example of a catalytic agent with a high conversion yield, approaching 100% (Maurice et al., 2000). Yet this catalyst is expensive to be used in large quantities for the components cooling in the engine. Alternatives exist, and Huang et al. (2002) showed that zeolite-based catalysts facilitate significant endothermic cooling with hydrocarbon fuels such as JP-7 and JP-8+100. Titanium-based reactors have also shown satisfactory catalytic effects at elevated temperatures with minimal coking (Siebenhaar et al., 1999).

Temperature and residence time both increase the quantity of solid deposits in the lines; the degree of conversion, however, reduces the amount of deposits. Coking can be mitigated by fuel deoxygenation or the use of additives that modify the reactions leading to the formation of heavy hydrocarbons and ultimately to long chains and deposits (Wickham et al., 1999); however, incorporation of these additives must be evaluated against the potential reduction of the fuel energy density or the possible negative effects on the reaction rates in the engine.

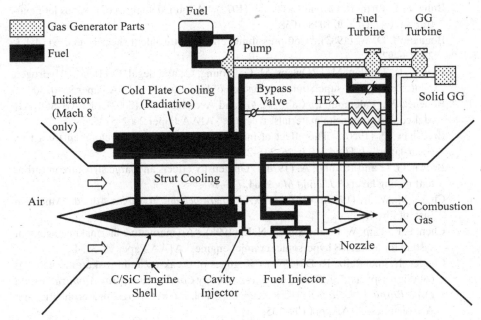

Figure 6.41. Fuel-cooling schematic (after Siebenhaar et al., 1999).

An example of a thermal management system is given in Fig. 6.41. The engine incorporates a strut-based fuel-injection configuration that, along with parts of the engine structure, is cooled by the fuel. Based on C/SiC components, cooling is provided by radiation to the surroundings, augmented by regeneratively cooled titanium structures. At Mach 8 conditions, the fuel cooling requires, in this design, 40% conversion from the endothermic fuel. The initiator shown in the figure is required to accelerate the endothermic reaction.

REFERENCES

Anderson, G. Y., McClinton, C. R., and Weidner, J. P. (2000). "Scramjet performance," in *Scramjet Propulsion* (E. T. Curran and S. N. B. Murthy, eds.), Vol. 189 of Progress in Astronautics and Aeronautics, AIAA, pp. 369–446.

Atsavapranee, P. and Gharib, M. (1997). "Structures in stratified plane mixing layers and the effects of cross-shear," *J. Fluid Mech.* **342**, 53–86.

Balakrishnan, G. and Williams, F. A. (1994). "Turbulent combustion regimes for hypersonic propulsion employing hydrogen-air diffusion flames," *J. Propul. Power* **10**, 434–437.

Balakrishnan, G., Smooke, M. D., and Williams, F. A. (1995). "A numerical investigation of extinction and ignition limits in laminar nonpremixed counterflowing hydrogen–air streams for both elementary and reduced chemistry," *Combust. Flame* **102**, 329–340.

Billig, F. S. (1988). "Combustion processes in supersonic flow," *J. Propul. Power* **4**, 209–216.

Billig, F. S. (1993). "Research on supersonic combustion," *J. Propul. Power* **9**, 499–514.

Billig, F. S., Orth, R. C., and Lasky, M. (1970). "A unified analysis of gaseous jet penetration," *AIAA J.* **9**, 1048–1058.

Bogdanoff, D. W. (1983). "Compressibility effects in turbulent shear layers," *AIAA J.* **23**, 926–927.

Bonghi, L., Dunlap, M. J., Owens, M. G., Young, C., and Segal, C. (1995). "Hydrogen piloted energy for supersonic combustion of liquid fuels," AIAA Paper 95–0730.

Bouchez, M., Falempin, F., Cahuzac, G., and Avrashkov, V. (2002). "PTAH-SOCAR fuel-cooled composite materials structure," AIAA Paper 2002–5135.

Bradshaw, P. (1966). "The effect of initial conditions on the development of a free shear-layer," *J. Fluid Mech.* **26**, 225–236.

Brown, G. L. and Roshko, A. (1974). "On density effects and large structure in turbulent mixing layers," *J. Fluid Mech.* **64**, 775–781.

Chase, M. W. Jr. (1998). *NIST-JANNAF Thermochemical Tables*, 4th ed., American Inst. of Physics.

Chen, F. F., Tam, W. F., and Shimp, N. R. (1998). "An innovative thermal management system for a Mach 8 hypersonic scramjet engine," AIAA Paper 98–3734.

Chigier, N. and Reitz, R. D. (1996). "Regimes of break-up and breakup mechanisms (physical aspects)," in *Recent Advances in Spray Combustion: Spray Atomization and Drop Burning Phenomena* (K. K. Kuo, ed.), Vol. 166 of Progress in Aeronautics and Astronautics, AIAA, pp. 109–135.

Chinzei, N., Masua, G., Komuro, T., Murakami, A., and Kudou, K. (1986). "Spreading of two-stream supersonic turbulent mixing layers," *Phys. Fluids* **29**, 1345–1347.

Chinzei, N., Mitani, T., and Yatsuyanagi, N. (2000). "Scramjet engine research at the National Aerospace Laboratory in Japan," in *Scramjet Propulsion* (E. T. Curran and S. N. B. Murthy, eds.), Vol. 189 of Progress in Astronautics and Aeronautics, AIAA, pp. 159–222.

Chomiak, J. (1990). *Combustion* (A. K. Gupta and D. G. Lilley, eds.), Energy and Engineering Science Series, Gordon and Breach.

Clemens, N. T. and Mungal, M. G. (1990). "Two- and three-dimensional effects in the supersonic mixing layer," AIAA Paper 90–1978.

Clemens, N. T. and Paul, P. H. (1995). "Scalar measurements in compressible axisymmetric mixing layers," *Phys. Fluids* **7**, 1071–1081.

Colket, M. B., III, and Spadaccini, L. J. (2001). "Scramjet autoignition study," *J. Propul. Power* **17**, 315–323.

Correa, S. M. and Mani, R. (1989). "Nonequilibrium model for hydrogen combustion in supersonic flow," *J. Propul. Power* **5**, 523–528.

Coordinating Research Council (CRC). (1988). *Aviation Fuel Properties*, Society of Automotive Engineers.

Curran, H. J., Gaffuri, P., Pitz, W. J., and Westbrook, C. K. (1998). "A comprehensive modeling study of *n*-heptane oxidation," *Combust. Flame* **114**, 149–177.

Dagault, P. and Cathonnet, M. (1998). "A comparative study of the kinetics of benzene formation from unsaturated C_2 to C_4 hydrocarbons," *Combust. Flame* **113**, 620–623.

Davidson, D. F., Horning, D. C., Herbon, J. T., and Hanson, R. K. (2001). "Shock tube measurements of JP-10 ignition," in *Proceedings of the Combustion Institute*, Vol. 28, pp. 1687–1692.

Day, M. J., Reynolds, W. C., and Mansour, N. N. (1998). "The structure of the compressible reacting mixing layer: Insights from linear stability analysis," *Phys. Fluids* **10**, 993–1007.

Dimotakis, P. E. (1986). "Two-dimensional shear-layer entrainment," *AIAA J.* **24**, 1791–1796.

Dimotakis, P. E. (1991). "Turbulent free shear layer mixing and combustion," *High-Speed Flight Propulsion Systems* (S. N. B. Murthy and E. T. Curran, eds.), Vol. 137 of Progress in Astronautics and Aeronautics, AIAA, pp. 265–340.

Diskin, C. S. and Northam, G. B. (1986). "Evaluation of a storable fluorine based pilot for scramjet," AIAA Paper 86–0372.

Drummond, J. P. (1997). "Enhancement of mixing and reaction in high-speed combustor flowfields," in *Advanced Computation and Analysis of Combustion* (G. D. Roy, S. M. Frolov, and P. Givi, eds.), ENAS Publishers, pp. 238–251.

Eggers, J. M. and Torrence, M. G. (1969). "An experimental investigation of the mixing of compressible-air jets in a coaxial configuration," NASA TN-D-5315.

Falempin, F. H. (2000). "Scramjet development in France," in *Scramjet Propulsion* (E. T. Curran and S. N. B. Murthy, eds.), Vol. 189 of Progress in Astronautics and Aeronautics, AIAA, pp. 47–118.

Ferri, A. (1973). "Mixing-controlled supersonic combustion," *Annu. Rev. Fluid Mech.* **5**, 301–338.

Fuller, R. P., Wu, P.-K., Kirkendall, K. A., and Nejad, A. S. (2000). "Effects of injection angle on atomization of liquid jets in transverse flow," *J. Propul. Power* **38**, 64–72.

Fuller, R. P., Wu, P.-K., Nejad, A. S., and Schetz, J. A. (1998). "Comparison of physical and aerodynamic ramps as fuel injectors in supersonic flow," *J. Propul. Power* **14**, 135–145.

Glassman, I. (1996). *Combustion*, 3rd ed., Academic Press.

Goldfeld, M. A., Mishunin, A. A., Starov, A. V., and Mathur, A. B. (2004). "Investigation of hydrocarbon fuels combustion in supersonic combustor," AIAA Paper 2004–3487.

Goldfeld, M. A., Starov, A. V., and Vinogradov, V. A. (2001). "Experimental study of scramjet module," *J. Propul. Power* **17**, 1222–1226.

Goltsev, V. F., Sverdlov, E. D., and Tulupov, Yu. N. (1991). "Combustion of methane flame in high velocity air flow with recirculation zone," in *Proceedings of the Third All-Union Workshop on Flame Structure*, Siberian Branch of the Russian Academy of Science, Novosibirsk, Russia, Vol. 1, pp. 147–151.

Gruenig, C., Avrashov, V., and Mayinger, F. (2000). "Fuel injection into a supersonic airflow by means of pylons," *J. Propul. Power* **16**, 29–34.

Guoskov, O. V., Kopchenov, V. I., Lomkov, K. E., Vinogradov, V. A., and Waltrup, P. J. (2001). "Numerical research of gaseous fuel preinjection in hypersonic three dimensional inlet," *J. Propul. Power* **17**, 1162–1169.

Hall, J. L., Dimotakis, P. E., and Rosemann, H. (1993). "Experiments in nonreacting compressible shear layers," *AIAA J.* **31**, 2247–2254.

Harradine, D. M., Lyman, J. L., Oldenborg, R. C., Schott, G. L., and Watanabe, H. H. (1990). "Hydrogen/air combustion calculations: The chemical basis of efficiency in hypersonic flows," *AIAA J.* **28**, 1740–1744.

Hartfield, R. J., Hollo, S. D., and McDaniel, J. C. (1994). "Experimental investigation of a swept ramp injector using laser-induced iodine fluorescence," *J. Propul. Power* **10**, 129–135.

Heiser, W. H. and Pratt, D. T. (1994). *Hypersonic Airbreathing Propulsion*, AIAA Education Series.

Hermanson, J. C. and Dimotakis, P. E. (1989). "Effects of heat release in a turbulent reacting shear layer," *J. Fluid Mech.* **159**, 151–168.

Hersch, M., Povinelli, L., and Povinelli, F. (1970). "A Schlieren technique for measuring jet penetration into a supersonic stream," *J. Spacecr. Rockets* **7**, 755–756.

Herzberg, G. (1989). *Molecular Spectra and Molecular Structure*, 2nd ed., Krieger, Vol. I.

Huang, H., Sobel, D. R., and Spadaccini, L. J. (2002). "Endothermic heat-sink of hydrocarbon fuels for scramjet cooling," AIAA Paper 2002–3871.

Huellmantel, L. W., Ziemer, R. W., and Cambel, A. B. (1957). "Stabilization of premixed propane-air flames in recessed ducts," *J. Jet Propul.* **27**, 31–43.

Hunt, J. L., Johnston, P. J., and Cubbage, J. M. (1982). "Hypersonic airbreathing missile concepts under study at Langley," AIAA Paper 82–0316.

Ianovski, L. S., Sosounov, V. A., and Shikhman, Yu. M. (1997). "The application of endothermic fuels for high speed propulsion systems," in *Proceedings of the 13th International Symposium on Air Breathing Engines*, ISABE 97–7007, pp. 59–69.

Inger, G. R. (2001). "Scaling nonequilibrium-reacting flows: The legacy of Gerhard Damköhler," *J. Spacecr. Rockets* **38**, 185–190.

Jachimowski, C. J. (1988). "An analytical study of the hydrogen–air reaction mechanism with application to scramjet combustion," NASA TP-2791.

Javoy, S., Naudet, V., Abid, S., and Paillard, C. E. (2003). "Elementary reaction kinetics studies of interest in H_2 supersonic combustion chemistry," *Exp. Thermal Fluid Sci.* **27**, 371–377.

Kanda, T., Haraiwa, T., Mitani, T., Tomioka, S., and Chinzei, N. (1997). "Mach 6 testing of a scramjet engine module," *J. Propul. Power* **13**, 543–551.

Kanda, T., Sunami, T., Tomioka, S., Tani, K., and Mitani, T. (2001). "Mach 8 testing of a scramjet engine module," *J. Propul. Power* **17**, 132–138.

Kanda, T., Tani, K., Kobayashi, K., Saito, T., and Sunami, T. (2002). "Mach 8 testing of a scramjet engine with ramp compression," *J. Propul. Power* **18**, 417–423.

Kay, I. W., Peschke, W. T., and Guile, R. N. (1990). "Hydrocarbon-fueled scramjet combustor investigation," AIAA Paper 90–2337.

Kopchenov, V. I. and Lomkov, K. E. (1992). "The enhancement of the mixing and combustion processes applied to scramjet engines," AIAA Paper 92–3428.

Kumar, A., Drummond, J. P., McClinton, C. R., and Hunt, J. L. (2001)."Research in hypersonic airbreathing propulsion at the NASA Langley Research Center," in *Proceedings of the 15th International Symposium on Air Breathing Engines*, AIAA.

Kush, E. A., Jr. and Schetz, J. A. (1973). "Liquid jet injection into a supersonic flow," *AIAA J.* **11**, 1223–1224.

Lander, H. and Nixon, A. C. (1971). "Endothermic fuels for hypersonic vehicles," *J. Aircr.* **8**, 200–207.

Leung, K. M. and Linstedt, R. P. (1995). "Detailed kinetic modeling of C_1–C_3 alkane diffusion flames," *Combust. Flame* **102**, 129–160.

Li, S. C., Varatharajan, B., and Williams, F. A. (2001). "Chemistry of JP-10 ignition," *AIAA J.* **39**, 2351–2356.

Li, S. C. and Williams, F. A. (1999). "NO_x formation in two-stage methane-air flames," *Combust. Flame* **118**, 399–414.

Linstedt, R. P. and Maurice, L. Q. (1995). "Detailed kinetic modeling of *n*-heptane combustion," *Combust. Sci. Technol.* **107**, 317–353.

Linstedt, R. P. and Maurice, L. Q. (2000). "Detailed chemical-kinetic model for aviation fuels," *J. Propul. Power* **16**, 187–195.

Liscinsky, D. S., True, B., and Holdeman, M. A. (1995). "Crossflow mixing of noncircular jets," AIAA Paper 95–0732.

Livingston, T., Segal, C., Schindler, M., and Vinogradov, V. A. (2000). "Penetration and spreading of liquid jets in an external–internal compression inlet," *AAIA J.* **38**, 989–994.

Marble, F. E., Zukoski, E. E., Jacobs, J. W., Hendricks, G. J., and Waitz, I. A. (1990). "Shock enhancement and control of hypersonic mixing and combustion," AIAA Paper 90–1981.

Masten, D. A., Hanson, R. K., and Bowman, C. T. (1990). "Shock-tube study of the reaction $H + O_2 \rightarrow OH + O$ using laser absorption," *J. Phys. Chem.* **94**, 7119–7128.

Maurice, L., Edwards, T., and Griffiths, J. (2000). "Liquid hydrocarbon fuels for hypersonic propulsion," in *Scramjet Propulsion* (E. T. Curran and S. N. B. Murthy, eds.), Vol. 189 of Progress in Astronautics and Aeronautics, AIAA, pp. 757–822.

McClinton, C. R. (1974). "Effect of ratio of wall boundary-layer thickness to jet diameter on mixing of a normal hydrogen jet in a supersonic stream," NASA TM X-3030.

McClinton, C. R. (2002). "Hypersonic technology . . . Past, present and future," inaugural presentation at the Institute for Future Space Transport, University of Florida.

Medwick, D. G., Castro, J. H., Sobel, D. R., Boyet, G., and Vidal, J. P. (1999). "Direct fuel cooled composite structure," ISABE 99–7284.

Mikolaitis, D. W., Segal, C., and Chandy, A. (2003). "Ignition delay for JP-10/air and JP-10/high energy density fuel/air mixtures," *J. Propul. Power* **19**, 601–606.

Mitani, T., Chinzei, N., and Kanda, T. (2001). "Reaction and mixing-controlled combustion in scramjet engines," *J. Propul. Power* **17**, 308–314.

Morrison, C. Q., Campbell, R. L., and Edelman, R. B. (1997). "Hydrocarbon fueled dual-mode ramjet/scramjet concept evaluation," in *Proceedings of ISABE 97*, Paper 97–7053, pp. 348–356.

Mungal, M. G., Hermanson, J. C., and Dimotakis, P. E. (1985). "Reynolds number effects on mixing and combustion in a reacting shear-layer," *AIAA J.* **23**, 1418–1423.

Murthy, S. N. B. (2000). "Basic performance assessment of scram combustors," in *Scramjet Propulsion* (E. T. Curran and S. N. B. Murthy, eds.), Vol. 189 of Progress in Astronautics and Aeronautics, AIAA, pp. 597–696.

Nishioka, M. and Law, C. K. (1997). "A study of ignition in the supersonic hydrogen/air laminar mixing layer," *Combust. Flame* **108**, 199–219.

Northam, G. B. and Anderson, G. Y. (1986). "Supersonic combustion ramjet research at Langley," AIAA Paper 89–0159.

Northam, G. B., Greenberg, I., Byington, C. S., and Capriotti, D. P. (1992). "Evaluation of parallel injector configurations for Mach 2 combustion," *J. Propul. Power* **8**, 491–499.

Ogorodnikov, D. A., Vinogradov, V. A., Shikhman, Yu. M., and Strokin, V. N. (1998). "Design and research Russian program of experimental hydrogen fueled dual mode scramjet: Choice of concept and results of pre-flight tests," AIAA Paper 98–1586.

Oran, E. S. and Boris, J. P. (2001). *Numerical Simulation of Reactive Flow*, 2nd ed., Cambridge University Press.

Orth, R. C., Schetz, J. A., and Billig, F. S. (1969). "The interaction and penetration of gaseous jets in supersonic flow," NASA CR-1386.

Ortwerth, P. J. (2000). "Scramjet flowpath integration," in *Scramjet Propulsion* (E. T. Curran and S. N. B. Murthy, eds.), Vol. 189 of Progress in Astronautics and Aeronautics, AIAA, pp. 1105–1293.

Ortwerth, P. J., Mathur, A. B., Segal, C., and Mullagiri, S. (1999). "Combustion stability limits of hydrogen in a non-premixed, supersonic flow," in *Proceedings of ISABE '99*, Paper 99–143.

Owens, M. G., Mullagiri, S., Segal, C., and Vinogradov, V. A. (2001a). "Effects of fuel pre-injection on mixing in a Mach 1.6 airflow," *J. Propul. Power* **17**, 605–610.

Owens, M. G., Mullagiri, S., Segal, C., Ortwerth, P. J., and Mathur, A. B. (2001b). "Thermal choking analyses in a supersonic combustor," *J. Propul. Power* **17**, 611–616.

Owens, M. G., Segal, C., and Auslander, A. H. (1997). "Effects of mixing schemes on kerosene combustion in a supersonic airstream," *J. Propul. Power* **13**, 525–531.

Owens, M. G., Tehranian, S., Segal, C., and Vinogradov, V. A. (1998). "Flameholding configurations for kerosene combustion in a Mach 1.8 airflow," *J. Propul. Power* **14**, 456–461.

Ozawa, R. I. (1971). "Survey of basic data on flame stabilization and propagation for high speed combustion systems," Marquardt Co., TR AFAPL-TR-70-81.

Papamoschou, D. and Roshko, A. (1986). "Observations of supersonic free shear layers," AIAA Paper 86–0162.

Papamoschou, D. and Roshko, A. (1988). "The compressible turbulent shear layer: An experimental study," *J. Fluid Mech.* **197**, 453–477.

Parent, B. and Sislian, J. P. (2003). "Effect of geometrical parameters on the mixing performance of cantilivered ramp injectors," *AIAA J.* **41**, 448–456.

Petullo, S. P. and Dolling, D. S. (1993). "Large-structure orientation in a compressible turbulent shear-layer," AIAA Paper 93–0545.

Pitz, R. W. and Daily, J. W. (1983). "Combustion in a turbulent mixing layer formed at a rearward-facing step," *AIAA J.* **21**, 1565–1570.

Poinsot, T. and Veynante, D. (2001). *Theoretical and Numerical Combustion*, R. T. Edwards, Inc.

Portz, R. and Segal, C. (2004). "Mixing in high-speed flows with thick boundary layers," AIAA Paper 2004–3655.

Povinelli, F. P. and Povinelli, L. A. (1971). "Correlation of secondary sonic and supersonic gaseous jet penetration into supersonic crossflows," NASA TN D-6370.

Pulsonetti, M. V. and Stalker, R. J. (1996). "A study of scramjet scaling," AIAA Paper 96–4533.

Ranzi, E., Gaffuri, P., Faravelli, T., and Dagaut, P. (1995). "A wide range modeling study of n-heptane oxidation," *Combust. Flame* **103**, 91–106.

Ramshaw, J. D. (1980). "Partial chemical equilibrium in fluid dynamics," *Phys. Fluids* **23**, 675–680.

Riggins, D. W. and McClinton, C. R. (1992)."A computational investigation of mixing and reacting flows in supersonic combustors," AIAA Paper 92–0626.

Riggins, D. W., McClinton, C. R., Rogers, R. C., and Bittner, R. D. (1995). "Investigation of scramjet strategies for high Mach number flows," *J. Propul. Power* **11**, 409–418.

Riggins, D. W. and Vitt, P. H. (1995). "Vortex generation and mixing in three-dimensional supersonic combustores," *J. Propul. Power* **11**, 419–426.

Rogers, R. C. (1971). "Mixing of hydrogen injected from multiple injectors normal to a supersonic airstream," NASA TN D-6476.

Rogers, R. C. and Schexnayder, C. J. (1981). "Chemical kinetics analysis of hydrogen-air ignition and reaction times," NASA TP-1856.

Rossmann, T., Mungal, M. G., and Hanson, R. K. (2000). "An experimental investigation of high-compressibility non-reacting mixing layers," AIAA Paper 2000–0663.

Sabel'nikov, V. A. and Prezin, V. I. (2000). "Scramjet research and development in Russia," in *Scramjet Propulsion* (E. T. Curran and S. N. B. Murthy, eds.), Vol. 189 of Progress in Astronautics and Aeronautics, AIAA, pp. 223–367.

Samimy, M. and Elliot, G. S. (1990). "Effects of compressibility on the characteristics of free shear flows," *AIAA J.* **28**, 439–445.

Samimy, M., Reeder, M. F., and Elliot, G. S. (1992). "Compressibility effects on large structures in free shear layers," *Phys. Fluids A* **4**, 1251–1258.

Sangiovanni, J. J., Barber, T. J., and Sayed, S. A. (1993). "Role of hydrogen/air chemistry in nozzle performance for a hypersonic propulsion system," *J. Propul. Power* **9**, 134–138.

Schetz, J. A. (1980). "Injection and mixing in turbulent flow," in Vol. 68 of Progress in Aeronautics and Astronautics, AIAA.

Schetz, J. A., and Billig, F. S. (1966). "Penetration of gaseous jets injected into a supersonic stream," *J. Spacecr. Rockets* **3**, 1658–1665.

Segal, C., Haj-Hariri, H., and McDaniel, J. C. (1995). "Effects of the chemical reaction model on calculations of supersonic combustion flows," *J. Propul. Power* **11**, 565–568.

Shchetinkov, E. S. (1973). "On piece-wise one-dimensional models of supersonic combustion," *Explosions Shock Waves* **9**, 409–417.

Siebenhaar, A., Chen, F. F., Karpuk, M., Hitch, B., and Edwards, T. (1999). "Engineering scale titanium endothermic fuel reactor demonstration for hypersonic scramjet engine," AIAA Paper 99–4909.

Sislian, J. P. (2000). "Detonation-wave ramjets," in *Scramjet Propulsion* (E. T. Curran and S. N. B. Murthy, eds.), Vol. 189 of Progress in Astronautics and Aeronautics, AIAA, pp. 823–890.

Slavinskaya, N. A. and Haidn, O. J. (2003). "Modeling of *n*-heptane and iso-octane gas-phase oxidation in air at low and high temperatures," AIAA Paper 2003–0662.

Smith, G. P., Golden, D. M., Frenklach, M., Moriarty, N. W., Eiteneer, B., Goldenberg, M., Bowman, C. T., Hanson, R. K., Song, S., Gardiner, W. C., Jr., Lissianski, V. V., and Qin, Z. (1999). "GRI Mech 3.0," available at http://www.me.berkeley.edu/gri_mech/.

Sobel, D. R. and Spadaccini, L. J. (1997). "Hydrocarbon fuel cooling for advanced propulsion," *ASME J. Eng. Gas Turbines Power* **119**, 344–351.

Starik, A. M. and Titova, N. S. (2001). "Initiation of combustion and detonation in $H_2 + O_2$ mixtures by excitation of electronic states of oxygen molecules," in *High-Speed Deflagration and Detonations: Fundamentals and Control* (G. D. Roy, S. M. Frolov, D. W. Netzer, and A. A. Borisov, eds.), ELEX-KM Pub.

Strokin, V. and Grachov, V. (1997). "The peculiarities of hydrogen combustion in model scramjet combustion," ISABE 97–7056, pp. 374–384.

Swithenbank, J., Eames, I., Chin, S., Ewan, B., Yang, Z., Cao, J., and Zhao, X. (1989). "Turbulence mixing in supersonic combustion systems," AIAA Paper 89–0260.

Thakur, A. and Segal, C. (2003). "Flameholding analyses in supersonic flow," AIAA Paper 2003–6909.

Varatharajan, B. and Williams, F. A. (2002a). "Ethylene ignition and detonation chemistry, part 1: Detailed modeling and experimental comparison," *J. Propul. Power* **18**, 344–351.

Varatharajan, B. and Williams, F. A. (2002b). "Ethylene ignition and detonation chemistry, part 2: Ignition histories and reduced mechanisms," *J. Propul. Power* **18**, 351–352.

Vasilev, V. I., Zakotenko, S. N., Krasheninnikov, S. J., and Stepanov, V. A. (1994). "Numerical investigations of mixing augmentation behind oblique shock waves," *AIAA J.* **32**, 311–316.

Vinogradov, V. A., Grachev, V., Petrov, M., and Sheechman, J. (1990). "Experimental investigation of 2-D dual mode scramjet with hydrogen fuel at Mach 4…6," AIAA Paper 90–5269.

Vinogradov, V. A. and Prudnikov, A. G. (1993). "Injection of liquid into the strut shadow at supersonic velocities," SAE-931455, Society of Automotive Engineers.

Vinogradov, V. A., Shikhman, Yu. M., and Segal, C. (2007). "A review of fuel pre-injection in supersonic, chemically reacting, flows," *ASME Appl. Mech. Revi.* **60**, 139–148.

Waltrup, P. J. (1987). "Liquid fueled supersonic combustion ramjets: A research perspective of the past, present and future," *J. Propul. Power* **3**, 515–524.

Waltrup, P. J. and Billig, F. S. (1973). "Structure of shock waves in cylindrical ducts," *AIAA J.* **11**, 1404–1408.

Warnatz, J., Maas, U., and Dibble, R. W. (1996). *Combustion*, Springer-Verlag.

Wendt, M., Stalker, R., and Jacobs, P. (1997). "Fuel stagnation temperature effects on mixing with supersonic combustion flows," *J. Propul. Power* **13**, 274–280.

Wickham, D. T., Ailtekin, G. O., Engel, J. R., and Karpuk, M. E. (1999). "Additives to reduce coking in endothermic heat exchangers," AIAA Paper 99–2215.

Williams, F. A. (1985). *Combustion Theory*, Addison-Wesley.

Williams, F. A. (1999). "Reduced kinetic schemes in combustion," in *Propulsion Combustion: Fuels to Emissions* (G. D. Roy, ed.), Taylor & Francis, pp. 93–128.

Wright, R. H. and Zukoski, E. E. (1960). " Flame spreading from bluff body flameholders," TR 34–17, JPL.

Yetter, R. A., Dryer, F. L., and Rabitz, H. A. (1991). "A comprehensive reaction mechanism for carbon monoxide/hydrogen/oxygen kinetics," *Combust. Sci. Technol.* **79**, 97–128.

7   Testing Methods and Wind Tunnels

7.1 Introduction

At Mach 5, the stagnation temperature approaches the structural limit acceptable for full simulation and continuous operation of a ground-based wind tunnel. With considerable effort, long-duration wind-tunnel operation could be
extended beyond this figure and, in fact, several capabilities exist in different places around the world. Experimental conditions above Mach 10 can
be duplicated only in short-duration facilities, so the topics studied in these
devices must be carefully selected to ensure that the physical processes to
be reproduced are compatible with those experienced in flight. Much of the
hypersonic flight domain remains to be covered through theoretical analysis
and, most likely limited, flight testing.

Despite the practical difficulties, considerable progress has been made
through experimental studies, in particular in the domain covering the transition from ramjet to scramjet operation, in the range of Mach 4–6. Beyond
this range, several short-duration flights were performed using boosters; basic
studies were undertaken in shock and expansion tubes at high Mach numbers.

The following is a brief summary, mostly of the hypersonic simulation
requirements and the capabilities offered by typical experimental facilities. Of
the numerous studies completed in ground facilities around the world, only
a few are referenced here as examples. For detailed reviews of current facilities and experimental programs, the volumes edited by Curran and Murthy
(2000) and Lu and Marren (2002), as well as the review by Arnold and Wendt
(1996) and the RAND report by Anton et al. (2004), are particularly recommended.

7.2 Hypersonic Flight Domain

The flight regimes that can be simulated in ground-based facilities become easily identified by examination of the hypersonic flight in an altitude–velocity

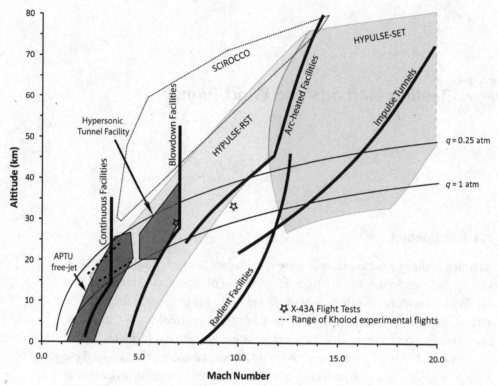

Figure 7.1. Predicted hypersonic flight range and the domain covered by existing facilities (Roudakov et al., 2001). APTU: Aerodynamic and Propulsion Test Unit at Arnold Engineering Development Center.

domain. Figure 7.1 shows a potential air-breathing corridor for a system including the scramjet cycle along with regions that have been reproduced in simulators. The air-breathing corridor is based on flight-path analyses that indicated that, for several optimized configurations (Escher, 1997), it is largely contained between dynamic pressure ratios of 0.25 and 1 atm (approx. 500–2000 psf). Imposed on the expected flight path are the capabilities of several types of ground-based facilities, including blowdown wind tunnels, with limited simulation capability and impulse facilities with short-time operation capable of simulating high Mach numbers. Also shown are some of the flight tests attempted to date, including the CIAM flights of 1992–1997; NASA's flights with X-43A ended in 2004. Not shown are some of the boosted flight tests from the Woomera Test Facility in South Australia that also exceeded Mach 8 during the scramjet operation.

Clearly, the ground-based facilities have limitations when required to fully simulate the flight conditions. This requires a careful analysis of the parameters that are, in fact, simulated during the ground-based studies to extract the correct relevance of the results for the eventual vehicle design and operation in flight.

The ground facilities included in the diagram in Fig. 7.1 mainly refer to blowdown facilities of long duration, arc-heated facilities that can also operate for a considerable experimental time, and shock tubes that along with impulse facilities operate for time sequences limited to a few seconds. In general, long-duration facilities simulate more of the airflow thermodynamic properties encountered during flight but cannot achieve the high Mach numbers that can be reproduced in short-duration facilities. In what follows, the principles of operations and some examples are described, limiting the discussion to applications to scramjet simulation and without reference to the requirements of external hypersonic aerodynamics.

7.3 Blowdown Facilities

Blowdown facilities contain an accumulated source of energy large enough to raise the air temperature to the desired flight enthalpy and use accelerating nozzles to increase the velocity to the experimental Mach number of interest. Then the model scramjet can be directly attached to the facility nozzle, in what is termed a direct-connect configuration, or inserted in the high-enthalpy stream. This becomes useful for inlet testing or combinations of inlet–isolator–combustion chamber flow paths. When the test duration is sufficiently long to affect the facility hardware's structural integrity, the maximum temperature – and hence the air enthalpy rise – becomes a limiting factor and the simulated flight Mach number has to be restricted. For this reason, long-duration blowdown facilities cannot exceed Mach 8 flight enthalpy. The high pressure encountered in flight is also difficult to simulate when approaching the high end of the flight Mach number, and it is often lower than the actual flight conditions; therefore the correct simulation of the flight Reynolds number along with temperature and velocity suffers. Several concepts exist to deliver the energy to the test airstream. Some of them are subsequently reviewed.

7.3.1 Combustion-Heated Wind Tunnels

A fuel, hydrogen or a hydrocarbon, is burned in the test air to raise the temperature to the desired level. The result is a vitiation of the facility's air with the resulting combustion products. An example of a vitiated wind tunnel is the NASA Langley Direct-Connect Supersonic Combustion Test Facility (DCSCTF) shown in Fig. 7.2. It is used to test combustor models in flows with stagnation enthalpies corresponding to flight at Mach numbers between 4 and 7.5. The DCSCTF (http://wte.larc.nasa.gov/facilities_updated/hypersonic/direct.htm) receives air from a high-pressure bottle field regulated to approximately 34 atm prior to entering the test cell. Gaseous hydrogen and oxygen are supplied from tube trailers with a maximum regulated pressure of

Figure 7.2. A model installed in the (DCSCTF) at NASA Langley's Scramjet Test Complex (http://wte.larc.nasa.gov/facilities_updated/hypersonic/direct.htm). The facility is used to test ramjet or scramjet combustors at conditions simulating flight Mach numbers from 4 to 7.5.

50 atm. Oxygen is used to replenish the amount consumed during air heating. A mixture of silane and hydrogen, 20%/80% by volume, is supplied in storage cylinders for use as an igniter of the primary fuel in the combustor models. Finally, stored nitrogen is available for purging residual gases at the end of an experiment. The facility is used for a range of studies, including mixing, ignition, flameholding, and combustion characteristics of the combustor models. The temperatures achievable in this facility range between 880 and 2100 K with mass flow rates in excess of 3 kg/s. The low temperature limit is dictated by the lower flammability limit of the vitiator, hydrogen in this case. The high temperature limit and the gases' storage capacity limit the experimental time to 20–30 s per run.

The vitiated heaters are compact, capable of relatively elevated temperatures as in the example just given, and, by comparison with other systems, relatively inexpensive. They are capable of a long operation time that is limited only by the size of the supply systems and the high temperatures tolerated by the materials used.

A major difficulty is caused by vitiation with combustion products that increase in proportion as the simulated temperature sought increases. For example, to achieve 1300 K, which simulates flight enthalpy corresponding approximately to Mach 5, the beginning of the hypersonic regime, as much as 13% mole fraction of water is generated in the test air if hydrogen is the heating compound. The presence of these combustion products, which do not exist in flight, has a significant impact on the chemical-reaction development in the model scramjet during testing. Water and carbon dioxide both have an effect on chain-terminating reactions; for example, the presence of water contributes to the formation of HO_2, an intermediate species that is sufficiently long lived at the relatively low temperature and pressure conditions of these simulations (Glassman, 1996). The result is a reduced rate of heat release when compared with similar but nonvitiated conditions; in addition, the presence of

Figure 7.3. Airflow reacting with hydrogen to generate a 1200 K test gas combined with hydrogen combustion with an equivalence ratio of 0.27 reduces the peak pressure rise by 9% higher compared with that of "dry" air conditions; only 5% water is present in the test gas in this case (Goyne et al., 2007).

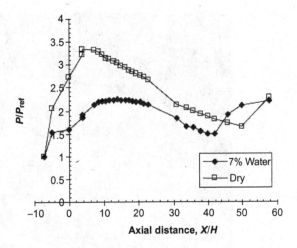

these species reduces the sensible energy released during the combustion process. In the study by Goyne et al. (2007) the reduced heat release was apparent from a lower level of pressure rise evidenced by the wall-measured pressure distribution, as shown in Fig. 7.3. Under conditions of Mach 2 airflow at 1200 K to react with hydrogen at an equivalence ratio of 0.27, the peak pressure rise for the nonvitiated case was 9% higher than that of the "dry" conditions. Only 5% water was added to the air in this case. Similarly, Tomioka et al. (2007) found the peak pressure lowered by 11% at stagnation temperatures of 1500 K when compared with that of the corresponding nonvitiated case. The effect becomes stronger as the experimental temperature has to increase. An alternative would be, evidently, a "clean" system based on electrical heating.

7.3.2 Electrically Heated Wind Tunnels

Electrical heating is highly desirable and difficult to achieve. First, a large source of energy must be made available. A facility of the scale of the DCSCTF described in Subsection 7.3.1 would require several megawatts of installed power. As the temperature increases, the heat transfer becomes less and less efficient, increasingly requiring more heating elements, thus leading to large-sized heaters. Krauss and McDaniel (1992) describe an electrically based facility that can be reliably and continuously operated in excess of 1250 K, i.e., close to a flight enthalpy of Mach 5 with maximum cold airflow rates of 0.75 kg/s. The facility is built by stacking heating stages, resulting in a heater assembly in excess of 8 m long.

Higher temperatures are difficult to obtain with existing heating elements, even when ceramic-based formulations are used. The temperatures experienced by the heating elements reach the domain when material softening occurs to the degree that requires vertical orientation; their heating efficiency

Figure 7.4. Aerial view of the HTF at NASA Glenn Research Center (http://facilities.
grc.nasa.gov). The nitrogen storage tank is visible in the lower part of the image, and
the oxygen tank farm on the central right side. The steam ejector is visible at the mid-
right section of the figure.

becomes, in this case, deficient, and their capability to contribute to an air temperature increase becomes problematic.

A particular solution for an electric-based heated facility is the Hypersonic
Tunnel Facility (HTF) at NASA Glenn Research Center (see Fig. 7.4), which
solved the issue of oxidative environment at elevated temperatures that would
affect regular heating elements. Clean nitrogen gas is heated through a 3-MW
graphite storage heater (Woike and Willis, 2002). Ambient-temperature oxygen is then mixed downstream of the heater to recreate air composition. Temperatures as high as 2170 K can be achieved with maximum 82 atm covering a
simulated flight range between 18 and 36 km with maximum Mach 7 enthalpy.
Run times range from 40 s to almost 5 min, depending on the desired flow
rate and temperature. The facility has a nozzle exit diameter in excess of 1 m,
enabling the testing of large-scale scramjet models. Complementing the facility's capability, three axisymmetric, interchangeable nozzles create exit conditions of Mach 5, 6, and 7. The model is suspended on the test chamber ceiling
and is introduced into the gas stream once the experimental conditions have
achieved the desired conditions. This is thus a "free-jet" operation, unlike
Langley's DCSCTF facility. Finally, a steam-based ejector reduces the pressure in the test chamber to simulate altitude.

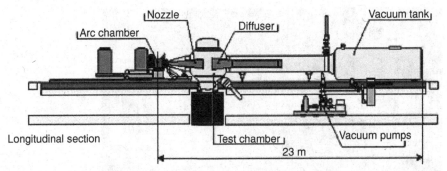

Figure 7.5. ONERA's F4 arc-heated wind tunnel (http://www.onera.fr/gmt-en/wind-tunnels/f4.php). An electric arc is applied to electrodes inserted in a stagnation chamber, leading to temperature and pressure rises. A plug is then open to blow the high-enthalpy gas over the test object. A vacuum tank complements the facility.

Hence the HTF is capable of large flow rates, long test runs, relatively elevated flight enthalpies, and accurate simulation of flight air composition within the operational window. To store the considerable energy required for testing in this facility, over 27 000 kg of graphite are stacked in the heater, which leads to considerable demands on the operational procedures. The heating process is long and may take as much as 100 h to achieve the elevated temperatures.

7.3.3 Arc-Heated Facilities

Elevated temperatures can be obtained in arc-heated wind tunnels. Flight temperatures in excess of Mach 8 can be simulated, with air reaching values of 3000 to 10 000 K (Smith et al., 2002). Geometrical configurations vary with the goal of stabilizing the arc column for controllable and safe operation. Run times can be quite long; some arc-heated facilities can operate up to 30 min, and mass flows can reach 8 kg/s for the shorter-duration operations. The installed power is considerable, exceeding tens of megawatts. Therefore facilities based on arc discharge require considerable effort and proximity to large electric sources.

An example of an arc-heated wind tunnel is ONERA's F4 impulse facility, shown in Fig. 7.5. It is based on a reservoir filled with cold gas at pressures up to 100 atm. The installed electric power is considerable; a generator delivers up to 150 MW to the electrodes for a short period of time, several tens of milliseconds. Once the stagnation conditions are reached, a plug is opened to the testing environment and a blowdown sequence lasts just under half of a second (Sagnier and Vérand, 1998). The maximum pressure can reach 500 atm, although, as in most arc-heated tunnels, variations in enthalpy and pressure from test to test are large (Arnold and Wendt, 1996).

Testing times are between 50 and 150 ms and Mach numbers as high as 16 can be reached. Although the testing times for these high enthalpies are quite

Figure 7.6. NASA Langley arc-heated wind tunnel. This facility is capable of simulating Mach 4.7–8. The 13-kW installed power results in stagnation temperatures up to 2800 K and pressures in excess of 40 atm.

large, the arc-heated wind tunnels are mostly appropriate for external aerodynamic testing rather than for scramjet studies. Electrode material is often released in the gas stream, and dissociation can lead to the formation of species that are not present in air during flight, most notably nitrogen oxides. The flow quality is difficult to infer and is largely dependent on the means used to stabilize the arc. Finally, the uncertainty in enthalpy and the difficulty of inferring it from the tunnel-measured parameters create difficulties in interpreting the results in these wind tunnels. Yet arc-heated wind tunnels have been used for scramjet testing, for example, the Hyper-X Mach 7 studies in the NASA Langley Arc-Heated Scramjet Facility (see Fig. 7.6) (Smith et al., 2002).

7.4 Short-Duration, Pulsed-Flow Wind Tunnels

This category includes devices that store energy that is later released for a short duration over the test object, resulting in large-Mach-number simulations. This group includes shock tubes connected to an accelerating nozzle, expansion tubes, and free-piston shock tubes. Examples of corresponding active facilities are given in the subsections that follow.

7.4.1 Shock Tunnels

A shock tunnel uses the energy stored in a high-pressure chamber, called the *driver tube*, to set a shock wave in motion when released into an adjacent chamber containing the *driven gas*. To increase the shock Mach number, a light gas is used as the driver gas. Further acceleration is obtained when the

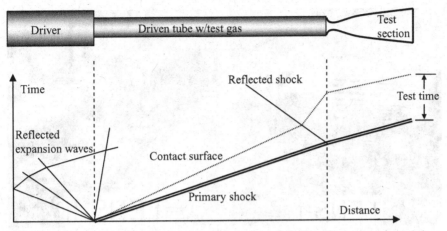

Figure 7.7. Space–time diagram of a shock-tube-based facility. The experimental time is dictated by the duration between the second diaphragm rupture and the arrival of the contact surface.

driver gas is heated. The initial pressure ratio between the driver and the driven gas increases rapidly as the desired shock Mach number grows; the pressure ratio depends on the selection of gases and their temperature, as indicated by Eq. 7.1. Usually a double diaphragm is used between the driver and the driven gas to improve the predictability of diaphragm rupture. At the end of the driven tube, a second diaphragm reflects the incident shock generating a stagnant, high-pressure, high-temperature volume of gas that, after the fracture of the second diaphragm, flows through an accelerating nozzle to establish the desired experimental conditions. The duration of steady experimental conditions is dictated by the time it takes the reflected shock to reach and interact with the driver–driven-gas contact surface. A space–time diagram of the shock-tunnel operation is shown in Fig. 7.7.

$$\frac{p_D}{p_d} = \frac{\gamma_D - 1}{\gamma_D + 1}\left[\frac{2\gamma_D}{\gamma_D - 1}M_s^2 - 1\right]\left[1 - \frac{\frac{\gamma_D - 1}{\gamma_d + 1}\left(\frac{a_D}{a_d}\right)(M_s^2 - 1)}{M_s}\right]^{-2\gamma_D/(\gamma_D - 1)}$$

$$(7.1)$$

Calspan facilities LENS I and LENS II shown in Fig. 7.8 are examples of shock-tube facilities.

LENS I has the capability to duplicate flight from Mach 7 to 14 (Holden and Parker, 2002); LENS II simulates flight at Mach 3–7. LENS I includes a 28-cm-diameter, 8-m-long, driver tube pressurized to over 2000 atm and electrically heated to drive a 20-cm-diameter, 20-m-long driven tube. The combination achieves a 6-ms-long test time at Mach 14 condition. Helium or hydrogen is used as the driver gas; the light, heated gas contributes to increase the shock Mach number when the pressure in the driver tube has reached the

Figure 7.8. Shock-tube facilities at Calspan-UB Research Center (CUBRC). LENS I (left) and LENS II (right) have capabilities of Mach 7–14 and 3–7, respectively (http://www.cubrc.org). Operational time at steady conditions ranges between 5 and 18 ms.

operational limit. A double-diaphragm solution ensures controllable rupture; the facilities are used with a combination of pressure and a mixture of gases that maximizes flow uniformity so that the contact surface has minimal interaction with the reflected shock: The contact surface is brought to rest. Finally, the contoured nozzles accelerate the flow to desired experimental conditions.

The Reynolds numbers achieved in the shock-tube facilities are large, exceeding 10^5 1/m and reaching as high as 10^8 1/m; thus flight conditions are better simulated than in the heated wind tunnels. The difficulty is the short duration that does not permit a full simulation of other important physical process, most notably heat transfer.

7.4.2 Free-Piston Shock Tubes

In a free-piston shock tube, the driven gas is compressed by a piston put in motion by the high pressure in a reservoir. The gas compressed by the piston plays the role of the drive gas in a shock tube. An example is the T5 facility at the Graduate Aeronautical Laboratories at the California Institute of Technology (GALCIT) shown in Fig. 7.9. Other similar facilities are operational at the University of Queensland in Australia (Paull and Stalker, 1998), at the German Aerospace Center in Göttingen, Germany (Hannemann and Beck, 2002), and elsewhere. It is named T5 because it is the fifth in a series of shock tunnels built by or under the supervision of R. J. Stalker from the University of Queensland in Australia. The facility is capable of producing flows of air or nitrogen up to a specific enthalpy of 25 MJ/kg, a pressure of 100 MPa, and a reservoir temperature of 10 000 K. It achieves this by using a free piston to adiabatically compress the driver gas of the shock tunnel to pressures as high as 1300 atm. The shock tube is 12 m long and 9 cm in diameter and is filled with

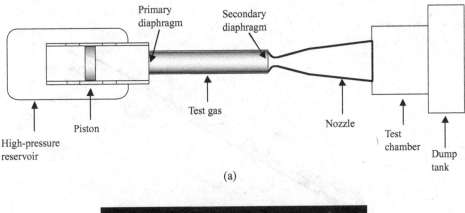

(a)

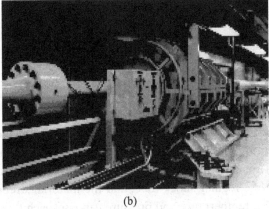

(b)

Figure 7.9. T5 free-piston tunnel at GALCIT.

a mixture of helium and argon. A thin diaphragm isolates the test gas from the test section and the dump tank from the gases in the test section. Both the test section and the dump tank are evacuated before a test.

The process starts with the release of high pressure upstream of the piston. The piston is accelerated to speeds in excess of 300 m/s, compressing the driver gas. When the primary diaphragm bursts, a shock wave propagates into the shock tube with a speed that can range from 2 to 5 km/s. The incident shock reflects off the end wall, the secondary diaphragm breaks, and the stationary gas behind reaches high temperature and pressure: the free-stream conditions at the nozzle exit are of the order of 0.3 atm and 2000 K temperature. Thus, although the temperature in a free-piston shock tube is usually large, the pressure is rather low. The T5 tunnel achieves 1–2 ms of steady operation before the pressure drops significantly.

7.4.3 Expansion Tubes

The expansion tube differs from the shock tube by the absence of the accelerating nozzle and the presence of an additional tube past the secondary

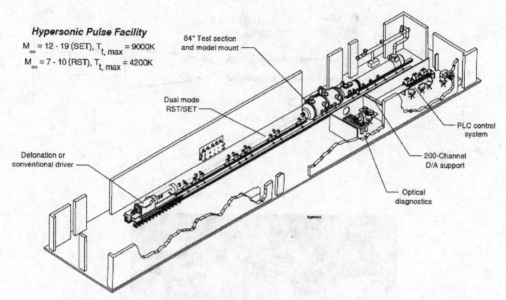

Hypersonic Pulse Facility
$M_\infty = 12 - 19$ (SET), $T_{t, max} = 9000K$
$M_\infty = 7 - 10$ (RST), $T_{t, max} = 4200K$

84" Test section and model mount

Dual mode RST/SET

Detonation or conventional driver

PLC control system

200-Channel D/A support

Optical diagnostics

Figure 7.10. The NASA HYPULSE expansion tube at GASL (Chue et al., 2002).

diaphragm, which allows the primary shock to penetrate and accelerate. The high temperature rise experienced behind the reflected shock in the usual shock tube is not experienced here. Consequently gas dissociation is not as significant as in the regular shock tube, and the material strength allows higher pressures; both temperature and pressure can approach those encountered in flight at high Mach. The test duration, however, suffers because it is limited to the time between the passage of the secondary shock wave and the arrival of the contact surface. Experimental times are reduced to less than 1 ms.

The NASA HYPULSE facility installed at GASL, shown in Fig. 7.10, accelerates air to the thermodynamic and kinematic conditions encountered in atmospheric flight from Mach 5 to 25. A thin diaphragm separates the shock tube from the acceleration tube section, which is open to the test chamber at near vacuum. An optional divergent nozzle that is directly connected to the expansion tube exit may also be used. The test gas is then accelerated in the expansion tube. Operation at higher pressures can be achieved with a shock-induced-detonation driver. In this operational mode, the initial shock passes through a combustible mixture in the driver-tube section, leading to the formation of a detonation shock and taking advantage of this added energy to drive the incident shock through the test gas at a higher speed. Hence, through a combination of methods to generate and accelerate the primary shock and the selection of working gases and expansion-tube geometry, the HYPULSE facility accommodates a broad range of flight conditions for scramjet and external aerodynamics testing. For example, Mach 15 flight enthalpy is simulated with 0.07 atm and 740 K in the test section; the experimental time is approximately 0.4 ms.

7.5 Summary

Duplicating flight conditions in ground facilities remains a difficult undertaking. Long-duration blowdown facilities are energy limited, and the maximum flight Mach number that can be expected does not exceed 8. Not all flight parameters can be accurately simulated, in particular the Reynolds number, thus limiting the accuracy of the flight simulation. Shock-tube tunnels can realize high enthalpies and can improve the simulation of all or most of the flight parameters up to orbital flight, but they are limited in the test duration so many necessary experimental studies, e.g., flameholding or heat transfer, cannot be performed in these facilities. Other solutions for testing exist, such as MHD accelerators or scaled models launched in ballistic facilities. Much of the development of the hypersonic vehicles will be based on theoretical analyses, yet many experimental programs in ground-based facilities will remain necessary to complement and validate the computational analyses and designs.

REFERENCES

Anton, P. S., Johnson, D. J., Block, M., Brown, M., Drezner, J., Dryden, J., Gritton, E. C., Hamilton, T., Hogan, T., Mesic, R., Peetz, D., Raman, R., Steinberg, P., Strong, J., and Trimble, W. (2004). "Wind tunnel and propulsion test facilities," RAND Corporation TR-134.

Arnold, J. and Wendt, J. F. (1996). "Test facilities," in *Hypersonic Experimental and Computational Capability, Improvement and Validation* (J. Muylaert, A. Kumar, and C. Dujarric, eds.), AGARD Advisory Report 319, pp. 174–200.

Chue, R. S. M., Tsia, C.-Y., Bakos, R. J., and Erdos, J. I. (2002). "NASA's HYPULSE facility at GASL – A dual mode, dual driver reflected-shock/expansion tunnel," in *Advanced Hypersonic Test Facilities* (F. K. Lu and D. E. Marren, eds.), Vol. 198 of Progress in Astronautics and Aeronautics, AIAA, pp. 29–71.

Curran, E. T. and Murthy, S. N. B. (eds.). (2000). *Scramjet Propulsion*, Vol. 189 of Progress in Astronautics and Aeronautics, AIAA.

Escher, W. J. D. (ed.). (1997). *The Synerget Engine: Airbreathing/Rocket Combined-Cycle Propulsion for Tomorrow's Space Transport*, Society of Automotive Engineers.

Glassman, I. (1996). *Combustion*, 3rd ed., Academic Press.

Goyne, C. P., McDaniel Jr., J. C., Krauss, R. H., and Whitehurst, W. B. (2007). "Test gas vitiation effects in a dual-mode scramjet combustor," *J. Propul. Power* **23**, 559–565.

Hannemann, K. and Beck, W. H. (2002). "Aerothermodynamics research at DLR High Enthalpy Shock Tunnel HEG," in *Advanced Hypersonic Test Facilities* (F. K. Lu and D. E. Marren, eds.), Vol. 198 of Progress in Astronautics and Aeronautics, AIAA, pp. 205–238.

Holden, M. S. and Parker, R. A. (2002) "LENS hypervelocity tunnels and application to vehicle testing at duplicated flight conditions," in *Advanced Hypersonic Test Facilities* (F. K. Lu and D. E. Marren, eds.), Vol. 198 of Progress in Astronautics and Aeronautics, AIAA, pp. 73–110.

Krauss, R. and McDaniel, J. C. (1992). "A clean air continuous flow propulsion facility," AIAA Paper 92–3912.

Lu, F. K. and Marren, D. E. (eds.). (2002). *Advanced Hypersonic Test Facilities*, Vol. 198 of Progress in Astronautics and Aeronautics, AIAA.

Paull, A. and Stalker, R. J. (1998). "Scramjet testing in the T4 impulse facility," AIAA Paper 98–1533.

Roudakov, A. S., Semenov, V. L., Strokin, M. V., Relin, V. L., Tsyplakov, V. V., and Kondratov, A. A. (2001). "The prospects of hypersonic engines in-flight testing technology development," AIAA-2001-1807, 10th International Space Planes and Hypersonic Systems, Kyoto, Japan.

Sagnier, P. and Vérand, J.-L. (1998). "Flow charaterization in ONERA F4 high-enthalpy wind tunnel," *AIAA J.* **36**, 522–531.

Smith, D. M., Felderman, E. J., and Shope, F. L. (2002). "Arc-heated facilities," in *Advanced Hypersonic Test Facilities* (F. K. Lu and D. E. Marren, eds.), Vol. 198 of Progress in Astronautics and Aeronautics, AIAA, pp. 279–314.

Tomioka, S., Hiraiwa, T., Kobayashi, K., and Izumikawa, M. (2007). "Vitiation effects on scramjet engine performance in Mach 6 flight conditions," *J. Propul. Power* **23**, 789–796.

Woike, M. R. and Willis, B. P. (2002). "NASA Glenn Research Center's Hypersonic Tunnel Facility," in *Advanced Hypersonic Test Facilities* (F. K. Lu and D. E. Marren, eds.), Vol. 198 of Progress in Astronautics and Aeronautics, AIAA, pp. 427–439.

8 Computational Fluid Dynamic Methods and Solutions for High-Speed Reacting Flows

8.1 Introduction

Possibly few propulsion systems can benefit more from the development and application of predictive tools than the scramjet engine does, and possibly few flows can be more challenging to model and simulate. It is encouraging therefore that, among the technology topics that are important for the scramjet engine development, the predictive capabilities of computational fluid dynamics have made some of the greatest advances over the past decade. Flow-field simulation, heat transfer, fluid–wall interaction treatment, incorporation of detailed chemical kinetics models, more efficient models and numerical schemes, computations, and algorithms have all evolved considerably and took advantage of the ever-increasing hardware capacity and speed.

The scramjet flow field presents the predictive tools with a particularly difficult environment. High-Reynolds-number flow regimes dominate this flow field with embedded regions of low speed in recirculation regions. Regions of large thermal and composition gradients are present along with complex shock-wave structures. Complex chemical reactions take place with large differences in time scales and species production. Heat release and flow interactions contribute to further increase the complexity of the simulation. Occasionally, additional flow features are present. Multiphase flows, for example, which are present when liquid fuels are used or when particulates form during the combustion process, add their new issues involving phase change, relative phase velocity, radiative transport, etc.

A detailed, temporally accurate description of the scramjet flow field is still a daunting task despite the progress made by the computational capacity and massive parallelization of present computational tools. The level of detail and the physical accuracy of the models used therefore must be traded in through various methods that facilitate sensible flow physics' traceability with reasonable computational efforts so their results can indeed contribute to technology advancement.

The topics facing modern reactive flow simulations are broad and cover all the aspects of numerical efficiency, implementation, and model accuracy. This chapter focuses on the modeling aspect of the physical processes present in the scramjet engine with emphasis on reactive flows. Issues related to numerical techniques, solution algorithms, or grid development have been the subjects of extensive specialized volumes and are not included here. References to these topics are made only in support of the modeling methods discussed in what follows.

8.2 Conservation Equations and Flow Physics Captured in These Equations

8.2.1 Field and Constitutive Equations

To solve a flow field, the system of equations must include the conservation of mass, momentum, energy, and species listed in Chap. 2, along with the constitutive equations. Chemically reacting flows also include the production–destruction of species, depending on the reaction rates included in the reaction model selected. These equations are briefly reproduced here.

- Mass conservation:

$$\frac{D}{Dt} \int_V \rho \, dV = 0. \tag{8.1}$$

- Momentum conservation:

$$\rho \frac{D\bar{u}}{Dt} = -\nabla p + \nabla \tilde{\tau} + \bar{F}_b. \tag{8.2}$$

- Energy conservation:

$$\rho \frac{D}{Dt} \left(e + \frac{u^2}{2} \right) = -\nabla (p \cdot \bar{u}) + \nabla (\tilde{\tau} \cdot \bar{u}) - \nabla \bar{q} + \bar{F}_b \cdot \bar{u} + Q. \tag{8.3}$$

- Species conservation in a multicomponent reacting mixture,

$$\frac{\partial Y_i}{\partial t} + \bar{u} \cdot \nabla Y_i = \frac{w_i}{\rho} - \nabla \cdot (Y_i V_{di}), \quad i = 1, \dots, N, \tag{8.4}$$

where Y_i are the mass fractions of species i. This formulation of species conservation – unlike Eq. (2.11) in Chap. 2 – now includes the diffusion velocities V_{di}.

The constitutive equations are then added with an appropriate equation of state,

$$p = p(\rho, T, N_i), \tag{8.5}$$

enthalpy,

$$h = h(T, p), \tag{8.6}$$

the Fourier law for heat transfer,

$$q_i = -k\frac{\partial T}{\partial x_i},\tag{8.7}$$

and the shear-stress tensor, which, for Newtonian fluids, takes the form

$$\tau_{ik} = \mu\left(\frac{\partial u_i}{\partial x_k} + \frac{\partial u_k}{\partial x_i}\right) - \frac{2}{3}\mu\left(\nabla\bar{u}\right)\delta_{ik}.\tag{8.8}$$

The generality of these flow equations must be complemented with terms that describe certain physical processes associated with convective and diffusive transport, including the molecular transport of species, thermal conductivity, and viscosity.

8.2.2 Molecular Transport of Species and Heat

Binary and Multicomponent Diffusion Coefficients
In a multispecies gas, species diffusion treatment leads to a complex equation that is substituted with simplified expressions based on the binary diffusion coefficient D_{ij}, derived for simple systems composed of only two species with negligible pressure gradients, through Fick's law (Williams, 1985):

$$D_{12} = -\frac{V_1 Y_1}{\nabla Y_1}.\tag{8.9}$$

Here, V_1 is the diffusion velocity of the first species in the binary system, where $Y_1 + Y_2 = 1$ and $Y_1 V_1 + Y_2 V_2 = 0$. The use of the binary diffusion coefficient is extended then to calculate the individual species diffusion coefficient in multispecies systems through the approximation of Hirshfelder and Curtiss (Hirshfelder et al., 1954) as

$$V_k X_k = -D_k \nabla X_k \quad\text{with}\quad D_k = \frac{1 - Y_k}{\displaystyle\sum_{j=1}^{N} X_j / D_{jk}}\tag{8.10}$$

where X_k are mole fractions of species k. The binary diffusion coefficients, in turn, are calculated by means of kinetic theory (Hirschfelder et al., 1954):

$$D_{12} = \frac{3kT}{4N\mu_{12}\sigma_{12}\bar{v}_{12}},\tag{8.11}$$

where k is Boltzmann's constant, N is the total number of molecules in the volume considered, μ_{12} is the reduced mass of the two molecules [$\mu_{12} = m_1 m_2 / (m_1 + m_2)$], σ_{12} is the collision cross section, and $\bar{v}_{12}$ is the average velocity in the Maxwellian distribution. Equation (8.11) shows that the binary diffusion coefficient increases with temperature and decreases with pressure and molecular weight.

Thermal diffusion, which manifests the tendency of lighter molecules to migrate toward regions of higher temperature, also known as the Soret effect, is usually one order of magnitude smaller than the species-gradient-based diffusion (Williams, 1985) and therefore is often neglected. However, in fluid systems that include hydrogen it may become a significant factor.

Thermal Conductivity

For a single-component gas, thermal conductivity is the proportionality constant between the heat flux and the temperature gradient:

$$\bar{q} = -\lambda \nabla T. \tag{8.12}$$

For monoatomic gases – or polyatomic gases with the internal modes of energy considered frozen – thermal conductivity is given by

$$\lambda_i^0 = \frac{8.322 \times 10^3}{\sigma_i^2 \Omega_{ii}^{(2.2)^*}} \left(\frac{T}{m_i} \right)^{1/2}, \tag{8.13}$$

where $\Omega_{ii}^{(2.2)^*}$ is a collisional integral normalized to its rigid sphere value (Hirschfelder et al., 1954). For polyatomic gases, λ_i is corrected through the Euken factor E_i:

$$\lambda_i = E_i \lambda_i^0 \quad \text{with} \quad E_i = 0.115 + 0.354 \frac{c_{p_i}}{k}. \tag{8.14}$$

E_i is derived as an average over all microscopic states, assuming that all states have the same diffusion coefficients, which for polar molecules may induce errors (Oran and Boris, 2001).

Thermal conductivity for mixtures of gases is then calculated with the Mason and Saxena (1958) equation:

$$\lambda_m = \sum_{i=1}^{N} \lambda_i \left[1 + \sum_{\substack{k=1 \\ k \neq i}}^{N} G_{ik} \frac{X_k}{X_i} \right]^{-1}, \tag{8.15}$$

$$G_{ik} = \frac{1.065}{2\sqrt{2}} \left(1 + \frac{m_i}{m_k} \right)^{-1/2} \left[1 + \left(\frac{\lambda_i^0}{\lambda_k^0} \right)^{1/2} \left(\frac{m_i}{m_k} \right)^{1/4} \right]^2. \tag{8.16}$$

In chemically reacting flows, thermal conductivity appears often through the thermal diffusivity term, $\alpha \equiv \lambda / \rho c_p$, which scales as T^a / p with the power exponent a between 1.5 and 2 (Williams, 1985).

Viscosity

The shear viscosity μ, which appears in the shear-stress tensor in Eq. (8.8), is found from kinetic theory (Hirschfelder et al., 1954) as

$$\mu = 5kT/8\Omega_{ii}^{(2.2)}. \tag{8.17}$$

For the gas mixture, Wilke's formula offers

$$\mu_m = \sum_{i=1}^{N} \mu_i \left[1 + \sum_{\substack{k=1 \\ k \neq i}}^{N} G'_{ik} \frac{X_k}{X_i} \right]^{-1}, \tag{8.18}$$

with

$$G'_{ik} = \frac{\sqrt{2}}{4} \left(1 + \frac{m_i}{m_k} \right)^{-1/2} \left[1 + \left(\frac{\mu_i}{\mu_k} \right)^{1/2} \left(\frac{m_i}{m_k} \right)^{1/4} \right]^2. \tag{8.19}$$

The viscosity scales with T^a, with a having a value between 0.5 and 1 (Williams, 1985).

Nondimensional Transport Coefficients

Often the field equations are used in a nondimensional fashion. In these cases the transport coefficients are nondimensionalized as well.

The Prandtl number Pr is defined as the relative importance of momentum and thermal diffusivity:

$$\text{Pr} \equiv \frac{\nu}{\lambda/(\rho c_p)} = \frac{\mu c_p}{\lambda}. \tag{8.20}$$

For diatomic gases with constant $\gamma = 1.4$, $\text{Pr} = 0.74$. As γ decreases, Pr approaches unity; it can be quite large for liquids.

The Lewis number Le captures the relative magnitude between the energy transported through conduction and diffusion:

$$\text{Le} \equiv \frac{\lambda}{\rho c_p D_{ij}}. \tag{8.21}$$

The Schmidt number Sc compares the momentum diffusion with mass diffusion,

$$\text{Sc} \equiv \frac{\mu}{\rho D_{ij}}, \tag{8.22}$$

and a relation among these three nondimensional numbers exists as

$$\text{Sc} = \text{Pr} \times \text{Le}. \tag{8.23}$$

Convective Transport

The motion of fluid parcels through space represents the convective transport and is defined in the equations of motion through the flux terms. Additional terms, which help clarify the physical processes, appear in fluid dynamic analyses. Of them, vorticity is a quantity that describes the rotational motion of parcels of fluid (Oran and Boris, 2001):

$$\bar{\omega} \equiv \nabla \times \bar{v}. \tag{8.24}$$

An equation for vorticity appears from the conservation of mass and momentum with the vorticity definition,

$$\frac{\partial \bar{\omega}}{\partial t} + \bar{\omega} \nabla \cdot \bar{v} = \bar{\omega} \cdot \nabla \bar{v} + \frac{\nabla \rho \times \nabla \left(p \bar{I} + \tilde{\tau} \right)}{\rho^2}, \tag{8.25}$$

where the second term on the right-hand side of the equation represents a vorticity source that is due to interactions between density and pressure gradients. If the fluid is irrotational, $\bar{\omega} = 0$, the flow is said to derive from a potential ϕ, namely, $\bar{v} = \nabla \phi$.

8.3 Turbulent Reacting Flow – Length Scales

The scramjet flow is, with the exception of localized regions, a high-Reynolds-number environment with dominant inertial effects so that viscous effects are not strong enough to quickly damp fluctuations; turbulence is therefore significant. Turbulence results from the onset of instabilities that are caused by differences in mechanical properties between adjacent flow regions (Hinze, 1959; Tennekes and Lumley, 1972; Pope, 2000): *Rayleigh–Taylor* instabilities derive from the acceleration of denser regions past a lighter fluid, as in buoyant flow. They interact with pressure differences and result in vorticity generation contributing to the second term on the right-hand side of Eq. (8.25). *Kelvin–Helmholtz* instabilities appear as a result of velocity differences, such as in shear flows originating from a splitter plate; they lead to the formation of coherent structures (Dimotakis, 1986). Additional interactions causing flow instabilities appear when the flows are chemically reacting and generate turbulence involving thermal, acoustic, and thermodiffusive interactions (Williams, 1985). Localized heat release through exothermic chemical reactions leads to volumetric expansion; the resulting density gradient contributes to generating additional vorticity through interaction with the existing pressure gradients (Oran and Boris, 2001). This vorticity is generated on the chemical-reaction scale and contributes to enhanced mixing. In turn, improved mixing contributes to accelerating the chemical reactions. This is a strong and complex coupling through which the energy of small flow scales helps mix and remove the flow inhomogeneities produced on a large scale (Oran and Boris, 2001).

If the turbulent fluctuations are considered homogeneous throughout the flow field under analysis, a statistical interpretation of the turbulence can be approached through the spectrum of the energy content, $e(k)$, associated with the velocity fluctuation wave number k, or its reciprocal turbulent eddy length scale, $1/k$ (Hinze, 1959). Because the energy spectrum is thus related exclusively to the turbulent scale, the assumption that turbulence is isotropic simplifies the flow analysis considerably and makes the computation of complex flows manageable. The methods by which turbulence is included in the computational model depend on the intended accuracy under the restrictions of the available computation capacity.

The energy density covers the range of sizes from the physical dimension of the system to the smallest scales at which dissipation of the kinetic energy into localized heat occurs. Along this range the turbulent energy cascades from the larger scale to the smaller scales with negligible energy transfer from the smaller scales to the larger scales (Oran and Boris, 2001). The *Kolmogorov* scale, l_k, at which energy dissipation occurs when viscous forces balance the inertial forces and therefore the Reynolds number Re_k, is unity. If the dissipation of the kinetic energy is defined as the turbulent energy $u'^2(k)$ divided by a time scale related to the velocity fluctuation wavelength,

$$\varepsilon = \frac{u'^2(k)}{k/u'} = \frac{u'^3}{k},$$
(8.26)

the smallest scales found in turbulent flow, l_k, are defined by the kinematic viscosity υ and the dissipation rate ε (Williams, 1985):

$$l_k = \left(\upsilon^3/\varepsilon\right)^{1/4}.$$
(8.27)

If the largest scale, the *integral scale l*, is assumed to be close to the characteristic size of the flow (Poinsot and Veynante, 2005), for example, the combustion chamber duct height, a relation between the Kolmogorov scale and the integral scale, is directly provided by the flow Reynolds number:

$$l_k = \frac{l}{Re^{3/4}},$$
(8.28)

which clearly shows that the Kolmogorov scale is considerably smaller than the characteristic flow size for high-Reynolds-number flows.

Between these limiting scales, the *Taylor scale* can be defined based on the magnitude of the *average* rate of viscous dissipation. If the velocity fluctuation used in Eq. (8.26) is used in the Reynolds number definition, then the Taylor length scale is found as

$$l_t = \frac{l}{\sqrt{Re_t}},$$
(8.29)

which clearly places the Taylor length scale between the integral scale and the Kolmogorov scale, $l \gg l_t \gg l_k$, for flows with large Reynolds number.

The vortical and turbulent structures within these ranges contain a substantial amount of the flow energy and reside for a long time. Whether assumed isotropic, steady or intermittent, independent or influenced by the interactions with the heat released through chemical reactions, the Navier–Stokes equations are considered inclusive of all the physical processes involved in the flow, and the methods selected to solve these equations under certain assumptions made with regard to the turbulent structures lead to different levels of accuracy, given the available computational resources.

8.4 Computational Approaches for Turbulent, Chemically Reacting Flows

If all the turbulent scales in the flow included between the integral scale, l, and the Kolmogorov scale, l_k, must be resolved, the full, time-accurate, Navier–Stokes equations have to be solved; no particular model that describes turbulence is needed. This *direct numerical simulation* (DNS) is certainly the most accurate representation of the flow but requires a level of computational effort that cannot be undertaken at this time, or even in the foreseeable future, as a solution for complex flows such as occur in the scramjet. The need to resolve all the scales in the flow and the large disparities between the time and the length scales when chemical reactions are present requires million-sized grids, even for volumes as small as cubic millimeters (Poinsot and Veynante, 2005). For these reasons, DNS remains a tool to study turbulence and, in some cases, to verify computations by other, simplified, methods as long as the Reynolds number remains at modest values of the order of ~100.

To simplify the computation the Navier–Stokes equations are replaced with averaged flow equations through techniques called *Reynolds-averaged Navier–Stokes* (RANS) in which averaged quantities of the instantaneous flow parameters are calculated. A turbulence model must be included in these calculations to handle the dynamic flow nature and *close* the computation. Additional equations are needed to describe chemical species production–destruction and heat generation in chemically reacting flows.

Between these two approaches lies the *large-eddy simulation* (LES), which explicitly solves only the large scales in the flow, leaving the small-scale effects to a model applied to *subgrid turbulence*. Low-frequency fluctuations of properties are captured by this method to the grid cutoff. Then the coupling between the modeled small scales and the computed large scales must be implemented.

The degree to which the time fluctuations are captured by these three methods is described in the diagram by Poinsot and Veynante (2005), reproduced in Fig. 8.1. Here, the temperature calculated at a given point in the turbulent flame shown on the left-hand side of the figure would be determined with different degrees of time accuracy by each of these methods. The RANS method provides an averaged value over the period of time under investigation; LES provides a time fluctuation at the frequency scale allowed by the grid-scale cutoff, and, although it contains some temporal information, it still filters the small fluctuations; the DNS would capture all fluctuations because turbulence is fully computed, rather than described by a model. In terms of the

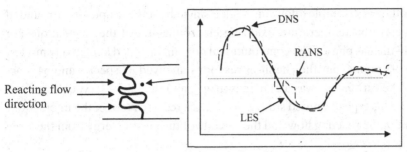

Figure 8.1. Time evolution of local temperature computed with DNS, LES, and RANS methods (after Poinsot and Veynante, 2005).

energy carried by the flow fluctuations, $e(k)$, LES is limited to those below the cutoff wave number, the RANS method provides all the energy range through the model that is adopted for the particular calculation, and DNS computes the entire range.

8.4.1 Direct Numerical Simulation

An example of DNS application to turbulent reacting flows by Wang and Rutland (2005) is shown in Fig. 8.2. It calculates a small region, 2 cm × 2 cm,

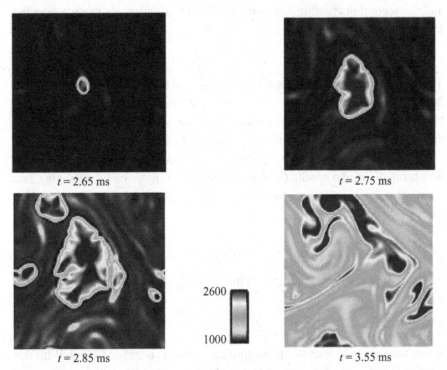

Figure 8.2. Temperature contours following ignition of a 35.5-μm droplet (Wang and Rutland, 2005).

of an otherwise complex flow, in which a single heptane droplet is surrounded by an elevated temperature oxidizing environment. All the calculations are done in the gas phase. The computational domain included a relatively modest 192×192 points, but the chemical reactions involved 33 species and 64 reactions. The turbulence was assumed isotropic and the initial Reynolds number was 50. This type of limited study is useful as a research tool of the interactions present in the reacting flow and the cascade of turbulent energy from the large to the small scale.

8.4.2 Reynolds-Averaged Navier–Stokes Simulation

In flows assumed steady in the mean the field variables are assumed to be amenable to decomposition into a mean and a fluctuating component. Following the formulation described by Oran and Boris (2001), this variable decomposition leads to

$$f = \bar{f} + f', \tag{8.30}$$

where the first component is averaged over a time interval Δt as

$$\bar{f} = \lim_{\Delta t \to \infty} \frac{1}{\Delta t} \int_t^{t+\Delta t} f(t) \, dt \tag{8.31}$$

and the second fluctuating component in Eq. (8.30) averages to zero over the same interval:

$$\overline{f'(x, t)} = 0. \tag{8.32}$$

This approach is correct under the assumption that variations in the mean flow are considerably slower than in the fluctuating terms (Oran and Boris, 2001); otherwise the two terms cannot be fully separated. Additional terms appear in the equations of motion (Williams, 1985), for example, the term $\overline{\rho' u_i'}$ in continuity equation (8.1), which represents a source term for the mean flow term, $\overline{\rho'} \cdot \overline{u_i'}$. To handle these additional terms, mass-based averaging, i.e., *Favre averaging*, has been suggested, with the averaged variables based on mass weighting:

$$\tilde{f} = \frac{\overline{\rho f}}{\bar{\rho}}. \tag{8.33}$$

The variables are now decomposed into a mean and a fluctuating component as

$$f = \tilde{f} + f'', \tag{8.34}$$

where the averaged fluctuating component amounts to zero, $\tilde{f}'' = 0$. With this approach and neglecting body forces and bulk heat addition, the equations of motion become as follows:

- Mass conservation:

$$\frac{\partial \rho}{\partial t} + \frac{\partial}{\partial x_i} (\bar{\rho} \tilde{u}_i) = 0. \tag{8.35}$$

- Momentum conservation:

$$\frac{\partial \bar{\rho} \tilde{u}_i}{\partial t} + \frac{\partial}{\partial x_i} (\bar{\rho} \tilde{u}_i \tilde{u}_j) = -\frac{\partial \bar{p}}{\partial x_j} + \frac{\partial}{\partial x_i} \left(\bar{\tau}_{ij} - \overline{\rho u_i'' u_j''} \right). \tag{8.36}$$

- Energy conservation:

$$\frac{\partial \bar{\rho} \tilde{h}}{\partial t} + \frac{\partial}{\partial x} (\bar{\rho} \tilde{u}_i \tilde{h}) = \frac{\partial \bar{p}}{\partial t} + \tilde{u}_i \frac{\partial \bar{p}}{\partial x_i} + \overline{u_i'' \frac{\partial p}{\partial x_i}}$$

$$+ \frac{\partial}{\partial x_i} \left(q_i - \overline{\rho u_i'' h''} \right) + \overline{\tau_{ij} \frac{\partial u_i''}{\partial x_j}} - \frac{\partial}{\partial x_i} \left(\rho \sum_{k=1}^{N} V_{ki} Y_k h_k \right), \tag{8.37}$$

where the last term derives from the chemical species with V_{ki} and Y_k describing the diffusion velocities and species concentration, respectively, as in Eq. (8.4), which, in turn, becomes in the Favre averaging formalism:

$$\frac{\partial \left(\bar{\rho} \tilde{Y}_k \right)}{\partial t} + \frac{\partial}{\partial x_i} (\bar{\rho} \tilde{u}_i \tilde{Y}_k) = -\frac{\partial}{\partial x_i} \left(\overline{V_{ki} Y_k} + \bar{\rho} \overline{u_i'' Y_k''} \right) + \bar{\dot{\omega}}_k \quad \text{for } k = 1, N. \tag{8.38}$$

In the incompressible limit, Favre-averaged equations become identical to Reynolds averaging. The averaged equations contain additional higher-order unknown terms such as the Reynolds stresses in the momentum equation and the turbulent heat fluxes in the energy equation. These terms require closure, which is done through turbulence models.

8.4.3 Turbulence Models

The Reynolds stress $\bar{\rho} \widetilde{u_i'' u_j''}$ is usually described for Newtonian fluids as a function of local velocity gradients through the turbulent dynamic viscosity μ_t and the turbulent kinetic energy $k = \frac{1}{2} \sum_{k=1}^{3} \widetilde{u_k'' u_k''}$ (Poinsot and Veynante, 2005):

$$\bar{\rho} \widetilde{u_i'' u_j''} = -\mu_t \left(\frac{\partial \tilde{u}_i}{\partial x_j} + \frac{\partial \tilde{u}_j}{\partial x_i} - \frac{2}{3} \delta_{ij} \frac{\partial \tilde{u}_k}{\partial x_k} \right) + \frac{2}{3} \bar{\rho} k. \tag{8.39}$$

In turn, the turbulent viscosity is determined as a function of local intensity and an appropriately selected length scale. These models vary from *zero-equation* models, which relate in an algebraic fashion the fluctuating quantities to the local mean flow, to *one-equation* models, which include an additional differential equation to calculate the turbulent kinetic energy and then relate the turbulent viscosity to this kinetic energy, to *two-equation* models in which the turbulent kinetic energy k and its dissipation rate ε, for example, are calculated separately by means of additional differential equations.

An example of a zero-equation model is the Prandtl model, which suggests that the turbulent viscosity quantity is related to the local velocity gradient, as in the example given by Poinsot and Veynante, (2005):

$$\mu_t = \bar{\rho} l_m^2 \tilde{S} \tag{8.40}$$

where $\tilde{S}$ is the mean stress tensor and l_m is an appropriately selected mixing length. This approach is useful for its simplicity, and the results are reasonably accurate for simple and well-characterized flows.

The one-equation model relates the turbulent viscosity to the turbulent kinetic energy,

$$\mu_t = \bar{\rho} C_\mu l_{pk} \sqrt{k} \tag{8.41}$$

where, as in the one-equation model, some calibrated constants, C and l_{pk}, are required. Here, the subscript "pk" in the mixing-length quantity identifies Eq. (8.40) as the Prandtl–Kolmogorov one-equation turbulence model.

The two-equation models include an additional equation to determine the length scale. In the $k - \varepsilon$ model of Jones and Launder (1972), for example, the turbulent viscosity is calculated as

$$\mu_t = \bar{\rho} C_\mu l_{pk} \frac{k^2}{\varepsilon}, \tag{8.42}$$

where two equations are used to determine both k and ε. These equations are coupled with the mean-velocity computation and rely on additional constants derived from prior calibration of experimental and computational analyses of the type of flow involved. The two-equation models assume isotropic turbulence, as the simpler models do, and their accuracy depends on the severity of property gradients away from the wall (Patel et al., 1985). These models perform the best for low-Reynolds-number flows.

More complex models exist, and some include higher-order computations; others account for nonisotropic turbulence and calculate the local turbulent quantities unrelated to wall functions. They introduce additional complexity into the computation. Oran and Boris (2001) include a detailed discussion of some of these models and refer to more detailed descriptions.

8.4.4 Large-Eddy Simulation (LES)

More detailed than RANS, LES methods compute the large turbulent structure in the flow and model the small-scale structures that are otherwise eliminated from the direct computation by filtering above a certain frequency and below a certain selected physical size. The flow features taking place at scales smaller than the applied filter are included through subgrid models that are then coupled to the computed solution of the larger scales. In the notation adopted by Poinsot and Veynante (2005), a filtered quantity f is averaged to a value $\bar{f}$, using a filter F, as

$$\bar{f}(x) = \int f(x')F(x - x')\, dx', \qquad (8.43)$$

where x is a vector or, possibly, a scalar quantity. Spatial filters can be of different forms, often including a low-pass, a Gaussian, or a cutoff filter in the spectral space, thus eliminating any structure below a selected physical size. As an example, a Gaussian filter in the physical space is

$$F(x) = F(x_1, x_2, x_3) = \left(\frac{6}{\pi \Delta^2}\right)^{\frac{3}{2}} \exp\left[-\frac{6}{\Delta^2}\left(x_1^2 + x_2^2 + x_3^2\right)\right]. \qquad (8.44)$$

Once selected, filters are normalized as

$$\int_{-\infty}^{\infty}\int_{-\infty}^{\infty}\int_{-\infty}^{\infty} F(x_1, x_2, x_3)\, dx_1 dx_2 dx_3 = 1. \qquad (8.45)$$

Through averaging, unclosed terms appear in the Navier–Stokes equations. They include a subgrid stress sensor, which is often solved with two-equation models such as, for example, the Smagorinsky model (Oran and Boris, 2001), which includes subgrid eddy viscosity and kinetic energy with dissipation determined from calibration experiments. In addition, subgrid terms that require closing include enthalpy flux, viscous work, convective and diffusive species, and heat fluxes (Sankaran et al., 2004). Further difficulties appear because of the interactions between the small-scale fluctuations, which are modeled, and the large scales, which are computed. These issues are discussed in detail by Oran and Boris (2001) and Poinsot and Veynante (2005) and treated extensively in numerous specific studies.

One particular interaction that appears particularly strong in scramjet flows is the energy exchange between the small scales at the subgrid level and the walls bounding the flow. Intense turbulence and large Reynolds numbers make the momentum exchange with the walls particularly important, especially in the context of computational model validation, which is often based on experimental results collected in small-sized facilities where the wall effects are considerable (Fureby et al., 2004).

8.5 Scramjet-Flow Computational Results

Along with continual improvement in computational capability in recent years, the simulation of scramjet flows has increased in detail and, arguably, in accuracy. A long list of problems has been treated computationally beginning with "unit problems," such as simplified jets, to complex flow geometries and flow interactions ranging from nonreacting steady flows in supersonic inlets to time-accurate, chemically reacting flows in combustion chambers of different

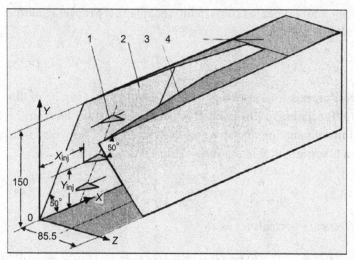

Figure 8.3. Schematic diagram of transverse injection behind pylons (1) installed in struts (2) placed vertically in the inlet. The geometry of the inlet walls (3) and cowl (4) was included in the computation (Gouskov et al., 2001).

geometries and transient simulations. A few examples from the large number of computations available are illustrated in the following subsections.

8.5.1 Steady-State Nonreacting Flows

The study by Gouskov et al. (2001) is an application of a 3D, Favre-averaged Navier–Stokes solver to a geometrical complex hypersonic inlet that includes fuel injection upstream of the isolator to increase the available mixing length. This computation was applied to the 3D scramjet model shown in Fig. 6.37 that was also subject to experimental testing. The computation was steady state and assumed a one-equation turbulence model. Multispecies mixing was included. The generic flow field, in which the fuel injectors particular to this solution were embedded, was computed with parabolized Navier–Stokes equations. The boundary-layer presence was eliminated based on previous analyses, which showed negligible effects of the boundary layers on the injection and the further development of the flow field in this problem.

The geometrical configuration used in this computation is shown in Fig. 8.3. The practical justification is based on the improved penetration and mixing when the fuel is injected transverse to the airflow behind thin pylons with thicknesses comparable to the jet-orifice diameter. Experimental evidence indicates that the drag induced by these types of pylons is negligible and their potential to lift the fuel from the wall surface, even at low injection pressures, is considerable (Livingston et al., 2000; Vinogradov et al., 2007).

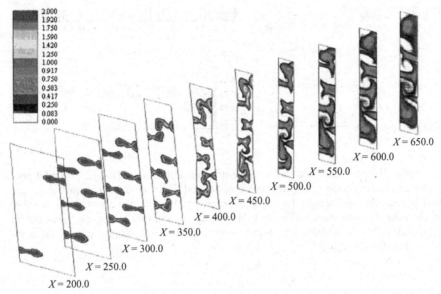

Figure 8.4. Equivalence ratio profiles for ethylene injection through six individual ports placed behind the pylons shown in Fig. 8.3. The overall equivalence ratio was 0.23.

This simplified analysis reduces the computational effort sufficiently to allow an optimization of the injection geometry for maximum fuel–air mixing efficiency. An example of the results is shown in Fig. 8.4 for six injection pylons placed on the two side walls with ethylene used as the injectant. The exit plane shows a relatively uniform fuel distribution with the equivalence ratio at any location dropping to one third of the injected values. Mixing efficiencies as high as 95%–98% were reported at the inlet discharge plane, as shown in Fig. 8.5, with overall equivalence ratios as high as 0.66. Clearly it is of great advantage to use the inlet length for mixing prior to arrival in the combustion chamber. The option and the practical implications raised by injection of part of the fuel in the inlet are discussed to a larger extent in Chap. 5.

Figure 8.5. Mixing efficiency for selected injection geometries. Pylon-based injection, discussed in Chap. 5, is shown in comparison with wall-based injection. The mixing efficiency was improved considerably in the course of this computation.

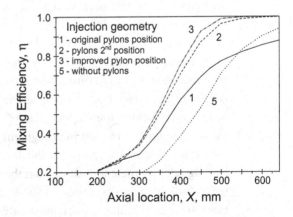

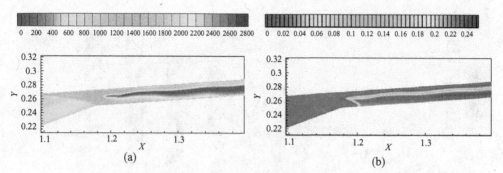

Figure 8.6. Temperature and water concentrations in the center plane. The inlet geometry and the injection configuration are optimized (i) to generate the shock structure that anchors the inlet terminal shock at the corner connecting the combustion chamber to the inlet, (ii) to ensure sufficient fuel–air mixing, and (iii) to generate a sufficient temperature rise by the shock system to initiate combustion immediately at the combustion chamber entrance.

8.5.2 Chemically Reacting Flows

Sislian and his collaborators focused computational studies (Parent and Sislian, 2004; Sislian and Parent, 2004; Sislian et al., 2006) on the shock-induced ignition concept. According to this design solution, the fuel is injected from the inlet ramp, mixes with the inlet-captured air, passes through the inlet shock system, and is finally ignited when the temperature raised by the shocks becomes sufficiently high.

The numerics solve Favre-averaged Navier–Stokes equations with a k–ω turbulence model for closure. The turbulent viscosity is calculated from $\mu_t = 0.09\rho k/\omega$ and is added to the molecular, diffusion-based viscosity. Thermal conductivity and mass diffusion are corrected with terms derived from this turbulent viscosity. Compressibility corrections are then introduced, given the high convective Mach number that characterizes the mixing processes. Finally, the chemical reactions are based on Jachimowski's model (Jachimowski, 1988) for hydrogen–air mixtures, which includes nine species and 20 reactions.

The physical model used by Sislian et al. (2006) includes a three-ramp inlet, cantilevered fuel injectors placed at the beginning of the second ramp, an additional slot injector extending the width of the inlet placed at the end of the second ramp, and a constant-area combustor attached to the inlet. A flight condition of Mach 11 at 35 km is assumed. Figure 8.6, reproduced from this computation, shows the computed temperature field and water production at the center plane of the last inlet segment and in the combustion chamber.

Noticeable in Fig. 8.6(a) is the last shock of the inlet anchored to the leading edge of the combustion chamber. Behind it, the temperature rises to sufficiently high values, around 1400 K, to ignite the mixture. At a short distance from the leading edge, considerable production of water is found in the

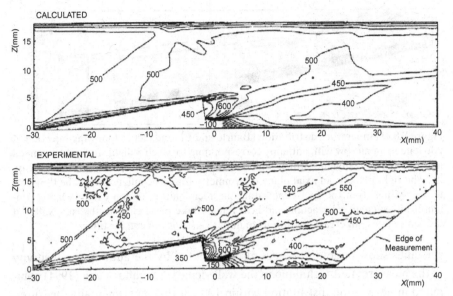

Figure 8.7. Velocity measured and computed at the center plane.

combustion chamber. The concept of using the inlet shock system to induce combustion of the mixture resulting from upstream injection – see the discussion of inlet fuel injection discussed in Chap. 5 – is proven, in principle, by these calculations; however, the operational stability under changing flight conditions, fuel throttling, regime transition from ramjet to scramjet, and low-enthalpy flight remains an issue.

An early comparison of computation with experiment was the study by Donahue et al. (1994), which included time-averaged velocity, temperature, pressure, and species concentration. The computation used SPARK, a family of Navier–Stokes equation solvers developed by Drummond (1988). It is based on a MacCormack predictor–corrector scheme with an algebraic Baldwin–Lomax turbulence model. For the jet–air mixing, a length was adopted based on the distance over which the jet concentration decayed to 5%. The flow field consisted of a tangential jet emitted from a swept ramp at Mach 1.7 into air flowing at Mach 2.0. In the experiment, planar laser iodine-induced fluorescence (PLIIF) was used to measure velocity from the Doppler frequency shift, species molar concentration from ratios of fluorescence collected with different seeding conditions, and pressure and temperature from the change in the fluorescence spectrum.

An example of a comparison between measured and computed parameters is shown in the centerline velocity distributions in Fig. 8.7. The computation captures quite well the velocity in the core flow but with larger differences noted in the recirculation region in the base of the ramp, possibly because of the simplicity of the algebraic turbulence model used here. The jet exit

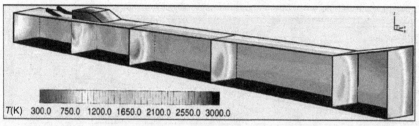

Figure 8.8. Temperature distribution (Baurle and Eklund, 2002) following injection of ethylene in an airflow with enthalpy corresponding to Mach 4 flight. Individual, angled injectors deliver ethylene, which ignites and sustains combustion by means of a cavity. The right plane is an axial plane of symmetry for this configuration. The flow pattern resulting from the fuel penetrating the recirculation region exhibits large non-uniformities; chemical reaction and heat release take place close to the side wall, but the center does not create conditions to initiate and sustain combustion.

velocities show large differences also attributed by the authors to sensitivity of the measurement technique in the wall vicinity. Donahue et al. (1994) computed mole-fraction distribution within 10% of the experimentally measured values, and the results suggest that the turbulence model used led to under-predicting the development of vortical motion on the ramp responsible for jet–air mixing. The same observation was made in the study by Riggins and Vitt (1995), who concluded that the algebraic model, as used for example in the SPARK codes, although offering an acceptable engineering mixing prediction, can indicate qualitatively only the downstream mixing without indicating the turbulent viscosity distribution.

The computation by Baurle and Eklund (2002) is an example that models an experiment at low Mach number in the range of 4–6 (Mathur et al., 1999). It includes mixing calculations upstream of the reaction zone and the development of chemical reactions in the combustion chamber. This regime is operationally difficult, representing the transition from ramjet to scramjet, and therefore it is of considerable interest. Under these conditions, the heat released through combustion increases the back pressure and leads to the formation of a shock train in the isolator that penetrates the combustion chamber. Large subsonic regions coexist along a supersonic core flow. Computationally these flows present difficulties because all regions must be solved simultaneously.

Baurle and Eklund (2002) used VULCAN, a RANS solver developed at NASA Langley with a high-Reynolds-number k–ω turbulence model near walls transitioning to the Jones–Launder k–ω model beyond boundary layers. For energy and mass transport calculations, the turbulent Prandtl and Schmidt numbers were assumed constant. Chemical reactions were modeled with a simplified three-step reaction mechanism for an ethylene–oxygen reaction. Figure 8.8 shows the temperature distribution in the combustion chamber from the Baurle and Eklund (2002) computation of the experiment performed at the U.S. Air Force Research Laboratory/Aerospace Propulsion Office (AFRL/PRA) at Mach 4 flight conditions. The configuration used four angled injectors inclined at 15° upstream of a cavity chosen as the flameholder; ethylene was the selected fuel. The figure shows half of the facility width with the right plane symmetrically dividing the flow. The calculation shows the corner of the cavity filled with combustion products whereas the center does not sustain the reactions. The region near the side wall indicates that reactions remain close to the wall, and the ensuing combustion products propagate downstream, further mixing and reacting with the incoming air. The authors attribute the proximity of the reactions to the wall to the slower flow in the region, and hence to increased residence time. Other factors can influence this reaction-region stratification; for example, the higher air temperature noted close to the wall at the inflow plane and the presence of additional recirculation regions – which were observed in other studies – in the vicinity of the step.

LES takes the computations further than the steady-state formulations and calculates the scales larger than the grid accurately both in time and space. The subgrid scales are then modeled. This technique was applied, for example, by Genin et al. (2003) to calculate a hydrogen-reacting jet emerging from a wedge placed in a Mach 2 flow. The subgrid modeling of the entire range of scales, including the effects of finite-rate chemistry and their interactions, is computationally difficult but cannot be neglected because the small scales dominate mixing and combustion. Genin et al. (2003) opt for a linear eddy mixing (LEM) model in conjunction with the LES calculation. The molecular diffusion and the chemical reactions leading to heat release are solved within the LEM at the computational cell size. The small-scale turbulence is assumed in this model to be locally homogeneous and isotropic and the heat release uniform. The subscale computation is thus 1D and, for accuracy, is selected to resolve scales down to the Kolmogorov scale. In this way, conservation of species does not need to be calculated at the LES scale because the chemical source term is resolved at the subgrid 1D scale; the 3D LES computation is responsible for large-scale convection. Thus, through this method, all the relevant physical processes are preserved; they are modeled separately but concurrently.

With this model and a reaction model involving two-step reactions for hydrogen (Rogers and Chinitz, 1982), the computation follows the conditions in the experiment by Guerra et al. (1991). Here, hydrogen emerges through three sonic injectors from the base of a ramp placed in a Mach 2 airstream. Figure 8.9 shows the geometry in the left diagram and a comparison of the measured and computed shadowgraphs on the right.

The right-hand figures compare the experimental shadowgraph image [Fig. 8.9(a)] with the computed shadowgraph [Fig. 8.9(b)]. The shocks and the wave structures are reasonably well captured by the computation, but the

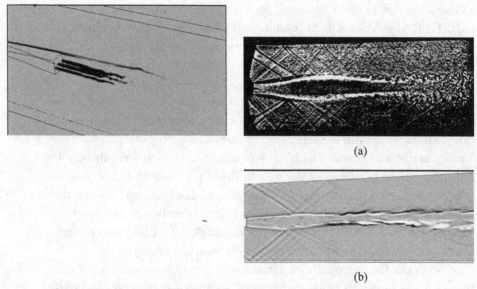

(a)

(b)

Figure 8.9. Left: hydrogen injection from sonic orifices into a Mach 2 flow and vorticity (Genin et al., 2003). Right: (a) experimental and (b) computational shadowgraphs for the reacting flow. The figures indicate that the computation captures well the shocks and the main flow structures; however, there is more expansion noted in the experiment than computed.

hydrogen jet expansion is underestimated. The effect cannot be attributed to reduced heat release in the computation because the simplified reaction mechanism would produce the opposite effect. The persistence of the jet core for a long distance after injection is made even clearer in the comparison of the nonreacting case, not included here. When compared with RANS calculations of the same experiment, the LES–LEM method showed improved agreement.

LES is a promising method for scramjet flow analyses. It has the versatility to use different subscale models within the same computation as, for example, was done in the study by Feiz and Menon (2003) to calculate unsteady, multiple jets injected in a cross flow. There, a finite-volume method used for the cross flow was coupled with a lattice Boltzmann equation computation of the jet before emerging from the nozzle. The model used within the LES framework remains to be validated for specific flows that depend on the dominant factors, such as the presence of compressible shear layers, large gradients in the proximity of walls, heat release and turbulence, phase interactions when liquids are present or when condensation occurs, etc.

8.6 Summary

Scramjet flows were consistently computed in recent years, covering all operational conditions and components. Models are improved to capture more physical processes, solvers become increasingly more efficient, and, along with them, computational power increases. Reaction mechanisms and flow interactions are improving in the simulations. It is difficult to predict when DNS will become a tool available for the complexities of high-speed, chemically reacting flows, but LES combined with advanced models continuously improves. As the understanding of these flows expands, the predictive capabilities offered by the development of computational simulations will become a primary tool for design, evaluation, and optimization of flight devices. It still is, and it will continue to be, a challenge to provide a broad experimental validation database because ground-based experimental facilities are energy limited and flight experiments are expensive. Yet experimental validation data are continuously acquired in a variety of wind tunnels covering many of the relevant issues. Finally, uncertainties of both experiment and computation must be carefully quantified to validate the numerical analyses with a comfortable degree of confidence.

REFERENCES

Baurle, R. A. and Eklund, D. R. (2002). "Analysis of dual-mode hydrocarbon scramjet operation at Mach 4–6.5," *J. Propul. Power* **18**, 990–1002.

Dimotakis, P. E. (1986). "Two-dimensional shear-layer entrainment," *AIAA J.* **24**, 1791–1796.

Donahue, J. M., McDaniel, J. C., and Hossein, H.-H. (1994). "Experimental and numerical study of swept ramp injection into a supersonic flowfield," *AIAA J.* **32**, 1860–1867.

Drummond, J. P. (1988). "A two dimensional numerical simulation of a supersonic, chemically reacting mixing layer," NASA TM-4055.

Feiz, H. and Menon, S. (2003). "LES of multiple jets in crossflow using a coupled lattice Boltzmann–Finite Volume Solver," AIAA Paper 2003–5206.

Fureby, C., Alin, N., Wikström, N., Menon, S., Svanstedt, N., and Persson, L. (2004). "Large eddy simulation of high-Reynolds-number wall-bounded flows," *AIAA J.* **42**, 457–468.

Genin, F., Chernyavski, B., and Menon, S. (2003). "Large eddy simulation of scramjet combustion using a subgrid mixing/combustion model," AIAA Paper 2003–7035.

Gouskov, O. V., Kopchenov, V. I., Lomkov, K. E., and Vinogradov, V. A. (2001). "Numerical research of gaseous fuel pre-injection in hypersonic three-dimensional inlet," *J. Propul. Power* **17**, 1162–1169.

Guerra, R., Waidmann, W., and Laible, C. (1991). "An experimental investigation of combustion of a hydrogen jet injected parallel in a supersonic airstream," AIAA Paper 1991–5102.

Hinze, J. O. (1959). *Turbulence – An Introduction to Its Mechanism and Theory*, McGraw-Hill.

Hirschfelder, J. O., Curtiss, C. F., and Bird, R. B. (1954). *Molecular Theory of Gases and Liquids*, Wiley.

Jachimowski, C. J. (1988). "An analytical study of the hydrogen–air reaction mechanism with application to scramjet combustion," NASA TP-2791.

Jones, W. P. and Launder, B. E. (1972). "The prediction of laminarization of a two-equation model of turbulence," *Int. J. Heat Mass Transfer* **15**, 301–314.

Livingston, T., Segal, C., Schindler, M., and Vinogradov, V. A. (2000). "Penetration and spreading of liquid jets in an external–internal compression inlet," *AIAA J.* **38**, 989–994.

Mason, E. A. and Saxena, S. C. (1958). "Approximate formula for the thermal conductivity of gas mixtures," *Phys. Fluids* **1**, 361–369.

Mathur, T., Streby, G., Gruber, M., Jackson, K., Donbar, J., Donaldson, W., Jackson, T., Smith, C., and Billig, F. (1999). "Supersonic combustion experiments with a cavity-based fuel injector," AIAA Paper 1999–2102.

Oran, E. S. and Boris, J. P. (2001). *Numerical Simulation of Reactive Flows.*, 2nd ed., Cambridge University Press.

Parent, B. and Sislian, J. P. (2004). "Validation of Wilcox k–ω model for flows characteristic to hypersonic airbreathing propulsion," *AIAA J.* **42**, 261–270.

Patel, V. C., Rodi, W., and Scheuerer, G. (1985). "Turbulence models for near-wall and low Reynolds number flows: A review," *AIAA J.* **23**, 1308–1319.

Poinsot, T. and Veynante, D. (2005). *Theoretical and Numerical Combustion*, 2nd ed., R. T. Edwards, Inc.

Pope, S. B. (2000). *Turbulent Flows*, Cambridge University Press.

Riggins, D. W. and Vitt, P. H. (1995). "Vortex generation and mixing in three-dimensional supersonic combustors," *J. Propul. Power* **11**, 419–426.

Rogers, R. C. and Chinitz, W. (1982). "On the use of hydrogen–air combustion model in the calculation of turbulent reacting flows," AIAA Paper 1982–0112.

Sankaran, V., Genin, F., and Menon, S. (2004). "A sub-grid mixing model for large eddy simulation of supersonic combustion," AIAA Paper 2004–0801.

Sislian, J. P. and Parent, B. (2004). "Hypervelocity fuel/air mixing in a shcramjet inlet," *J. Propul. Power* **20**, 263–272.

Sislian, J. P., Martens, R. P., Schwartzentruber, T. E., and Parent, B. (2006). "Numerical simulation of a real shcramjet flowfield," *J. Propul. Power* **22**, 1039–1048.

Vinogradov, V. A., Shikhman, Yu. M., and Segal, C. (2007). "A review of fuel pre-injection in supersonic, chemically reacting, flows," *ASME Appl. Mech. Rev.* **60**, 139–148.

Tennekes, H. and Lumley, J. L. (1972). *A First Course in Turbulence*, MIT Press.

Wang, Y. and Rutland, C. J. (2005). "DNS study of the ignition of n-heptane fuel spray under high pressure and lean conditions," *J. Phys.: Conference Series* **16**, 124–128.

Williams, F. A. (1985). *Combustion Theory*, Addison-Wesley.

Index

Index

Printed in the United States
By Bookmasters